GEOGRAPHY

FOR EDEXCEL

A LEVEL YEAR 1 AND AS

REVISION GUIDE

Series editor

Bob Digby

Catherine Hurst

Rebecca Tudor

OXFORD
UNIVERSITY PRESS

OXFORD
UNIVERSITY PRESS

Great Clarendon Street, Oxford, OX2 6DP, United Kingdom

Oxford University Press is a department of the University of Oxford.
It furthers the University's objective of excellence in research, scholarship,
and education by publishing worldwide. Oxford is a registered trade mark
of Oxford University Press in the UK and in certain other countries

© Oxford University Press 2018

Series editor: Bob Digby

Authors: Catherine Hurst, Rebecca Tudor

The moral rights of the authors have been asserted.

Database right of Oxford University Press (maker) 2018.

First published in 2018

British Library Cataloguing in Publication Data
Data available

ISBN 978-019-843272-2

10 9 8 7 6

Paper used in the production of this book is a natural, recyclable product made
from wood grown in sustainable forests. The manufacturing process conforms
to the environmental regulations of the country of origin.

Printed by CPI Group (UK) Ltd, Croydon CR0 4YY

Acknowledgements
The publisher and authors would like to thank the following for permission
to use photographs and other copyright material:

Cover: Eric Isselee/Shutterstock; **p10:** Bob Digby; **p17:** Diego Azubel/Epa/
REX/Shutterstock; **p22:** NASA's MODIS Rapid Response Team; **p23:** AFP/
Getty Images; **p24:** YOMIURI SHIMBUN/Getty Images; **p25(l):** Digital Globe;
p25(r): Digital Globe; **p35(t):** APICHART WEERAWONG/REX/Shutterstock;
p35(b): AAMIR QURESHI/Getty Images; **p41:** Canadian Parks Agency; **p48:**
Photograph courtesy from www.markhewittphotography; **p49:** funkyfood
London - Paul Williams/Alamy Stock Photo; **p53:** Bob Digby; **p69(t):** DAVE
CAULKIN/AP/REX/Shutterstock; **p69(b):** Doru Cristache/Shutterstock; **p72(t):**
Catherine Hurst; **p72(b):** frans lemmens/Alamy Stock Photo; **p87:** TED
ALJIBE/Getty Images; **p90:** Henry Westheim Photography/Alamy Stock Photo;
p91: Andrew Melbourne/Alamy Stock Photo; **p93(t):** Jon Wilson/Alamy Stock
Photo; **p93(b):** MARWAN NAAMANI/Getty Images; **p94:** Joe Giddens/PA
Archive/PA Images; **p120(t):** Jeff Gilbert/Alamy Stock Photo; **p120(b):** Andy
Buchanan/Alamy Stock Photo; **p121:** Courtesy of Urban Splash; **p124:** Bob
Digby; **p125:** Loop Images Ltd/Alamy Stock Photo; **p126:** Superfast Cornwall;
p127(t): Simon Burt/Apex News & Pictures; **p127(b), 153:** Bob Digby; **p103(t):**
Ricker Potter; **p103(b):** A C Manley/Shutterstock; **p104, 106:** Bob Digby;
p108: Frederick Wilfred; **p109:** Universal Images Group/Getty Images;
p110:, 112 Bob Digby; **p115(l):** Hallsville Quarter development image-Aecom;
p115(r): Bob Digby; **p134:** Leo Rosser/Alamy Stock Photo; **p136:** Courtesy of
Manchester Libraries, Information and Archives, Manchester; **p138, 139,
140:** Bob Digby; **p143:** Homer Sykes Archive/Alamy Stock Photo; **p144:** Cath
Harries/Alamy Stock Photo; **p146:** Bob Digby; **p147:** Loop Images Ltd/Alamy
Stock Photo; **p152:** Bob Digby.

Artwork by Dave Russell, Simon Tegg, Mike Parsons (Barking Dog),
Mike Connor, Lovell Johns, and Kamae Design.

Every effort has been made to contact copyright holders of material
reproduced in this book. Any omissions will be rectified in subsequent
printings if notice is given to the publisher.

Contents

Contents

Introduction: Aiming for success

If you want to be successful in your exams, then you need to revise all you've learned for your A level course! That can seem daunting – but it's why this book has been written. It contains key points that you need to learn to prepare for exams for the Edexcel A Level Geography specification.

This Revision Guide is one of five publications from Oxford University Press to support your learning. The others are:

- *Geography for Edexcel A Level Year 1 and AS* student book, which this Revision Guide works alongside. All page links in this book refer to the student book.
- *Geography for Edexcel A Level Year 2* student book, for which there is also a separate Revision Guide.
- *A Level Geography for Edexcel Exam Practice*.

How to use this book

This Revision Guide contains the following to help you revise for the Edexcel A Level Geography specification.

An introduction to each of Papers 1–3

These contain outlines of:

- the three exam papers you'll be taking (pages 6–7)
- the key topics from the specification in each Paper.

Content summaries

Each section in this Revision Guide summarises exactly the corresponding section in the student book, thus covering all the topics in the specification. Key content in the student book is summarised:

- each two-page topic in the student book is summarised on a single page in this Revision Guide
- each four-page topic is summarised on two pages.

Each section contains the following features:

You need to know – at the start of every section, this feature summarises key things you need to learn for each topic.

Main content – a summary of the main content found in the student book.

Ten-second summary – this summarises the essentials that you need to know, like a checklist.

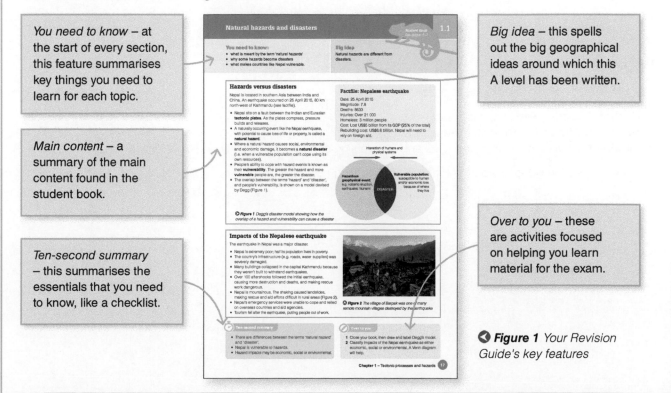

Big idea – this spells out the big geographical ideas around which this A level has been written.

Over to you – these are activities focused on helping you learn material for the exam.

◀ **Figure 1** *Your Revision Guide's key features*

Topics and question types on each exam paper

The Edexcel A Level Geography specification has eight topics, which together are assessed by three exams (Papers 1–3). Broadly, topics on physical geography are on Paper 1 and topics on human geography are on Paper 2. Paper 3 is a synoptic exam, designed to draw parts of the course together.

Paper 1

This paper has three sections, each assessing particular topics.

- **Section A** Tectonic Processes and Hazards
- **Section B Either** Glaciated Landscapes and Change **or** Coastal Landscapes and Change
- **Section C** The Water Cycle and Water Insecurity, **and** The Carbon Cycle and Energy Security

Marks and question types for each section are shown in Figure 2.

Section	Marks	Topics	Question	Details
A	16	Tectonic Processes and Hazards	Q1	• 1 x 4-mark question assessing quantitative skills • 1 x 12-mark essay question using the command word 'Assess'
B	40	**Either** Glaciated Landscapes and Change **or** Coastal Landscapes and Change	**Either** Q2 **or** Q3	• 2 x 6-mark paragraph questions using the command word 'Explain' • 1 x 8-mark longer paragraph question using the command word 'Explain' • 1 x 20-mark essay question using the command word 'Evaluate'
C	49	The Water Cycle and Water Insecurity **and** The Carbon Cycle and Energy Security	Q4	• 1 x 3-mark short paragraph question using the command word 'Explain' • 1 x 6-mark paragraph question using the command word 'Explain' • 1 x 8-mark longer paragraph question using the command word 'Explain' • 1 x 12-mark essay question using the command word 'Assess' • 1 x 20-mark essay question using the command word 'Evaluate'

🔺 **Figure 2** *Paper 1 topics, question styles and command words*

Paper 2

This paper also has three sections, each assessing particular topics.

- **Section A** Globalisation and Superpowers
- **Section B Either** Regenerating Places **or** Diverse Places
- **Section C Either** Health, Human Rights and Intervention **or** Migration, Identity and Sovereignty

Marks and question types for each section are shown in Figure 3.

Section	Marks	Topics	Question	Details
A	32	Globalisation **and** Superpowers	Q1 **and** Q2	For Globalisation • **Either** 1 x 4-mark short paragraph question **or** 1 x 4 marks assessing quantitative skills • 1 x 12-mark essay question using the command word 'Assess' For Superpowers • **Either** 1 x 4-mark short paragraph question **or** 1 x 4 marks assessing quantitative skills • 1 x 12-mark essay question using the command word 'Assess'
B	35	**Either** Regenerating Places **or** Diverse Places	**Either** Q3 **or** Q4	• 1 x 3-mark short paragraph question using the command word 'Suggest' • 2 x 6-mark paragraph questions using the command words 'Suggest' and 'Explain' • 1 x 20-mark essay question using the command word 'Evaluate'
C	38	**Either** Health, Human Rights and Intervention **or** Migration, Identity and Sovereignty	**Either** Q5 **or** Q6	• **Either** 1 x 4 mark short paragraph question **or** 4 marks for quantitative skills • 1 x 6-mark paragraph question using the command word 'Explain' • 1 x 8-mark longer paragraph question using the command word 'Explain' • 1 x 20-mark essay question using the command word 'Evaluate'

▲ **Figure 3** Paper 2 topics, question styles and command words

Paper 3

Paper 3 is a synoptic paper in the form of an issue analysis, presented as an unseen Resource Booklet on which questions are set. The topic will be drawn from the five compulsory topics in the specification, i.e.

• Tectonic Hazards (Topic 1)
• Globalisation (Topic 3)
• The Water Cycle and Water Insecurity (Topic 5)
• The Carbon Cycle and Energy Security (Topic 6)
• Superpowers (Topic 7).

Marks and question types for each section are shown in Figure 4.

Section	Marks	Question	Details
A	12	Q1–3	• 1 x 4-mark short paragraph question using the command word 'Explain' • 1 x 4-mark question assessing quantitative skills • 1 x 4-mark short paragraph question using the command words 'Explain' or 'Suggest'.
B	16	Q4–5	• 2 x 8-mark longer paragraph questions using the command word 'Analyse' (based on data in the Resource Booklet)
C	42	Q6–7	• 1 x 18-mark essay question using the command word 'Evaluate' (based largely on the Resource Booklet) • 1 x 24-mark essay question using the command word 'Evaluate' (based on the Resource Booklet in a wider context)

▲ **Figure 4** Paper 3 question styles and command words

Command words and marks

In order to assess students of different abilities, examiners use an 'incline of difficulty' in each exam paper. This means the questions early in each section count for fewer marks and are more straightforward than those that appear later with higher marks.

To aid this, examiners use the command words shown in Figure 5. Each carries a certain number of marks, as shown in Figure 6. Some command words are designed to be more challenging than others. Command words are used consistently throughout the three exam papers, so a question using the command word 'Assess' carries 12 marks, whether it is in Paper 1 or Paper 2.

	Command word	Definition
Low-tariff questions	Calculate	Produce a numerical answer, using relevant working.
	Draw/Plot	Create a graphical representation of geographical information.
	Complete	Create a graphical representation of geographical information by adding detail to a resource that's provided.
	Suggest	For an unfamiliar scenario, provide a reasoned explanation of how or why something may occur.
	Explain	Provide a reasoned explanation of how or why something occurs. An explanation requires a justification/exemplification of a point.
Medium-tariff questions	Analyse (only used in Paper 3)	Break something down into individual components/processes and say how each contributes to the question's theme/topic and how components/processes work together.
High-tariff questions	Assess	Use evidence to determine the relative significance of something. Give balanced consideration to all factors and identify which are the most important.
	Evaluate	Measure the value or success of something and ultimately provide a balanced and substantiated judgement/conclusion. Review information and bring it together to form a conclusion, drawing on evidence, e.g. strengths, weaknesses, alternatives and relevant data.

🔺 **Figure 5** *Command words used in Edexcel A Level Geography exams*

Command word	3	4	6	8	12	18	20	24
Calculate		*						
Draw/Plot/Complete		*						
Suggest	*	*	*					
Explain	*	*	*	*				
Analyse (Paper 3 only)				*				
Assess					*			
Evaluate						* Paper 3 only	* Papers 1 and 2 only	* Paper 3 only

🔺 **Figure 6** *Marks used for each command word*

Although the command words are used consistently across the exam papers, there are some slight variations.

- **Calculate** – normally requires a process of calculation, with marks awarded for the process of reaching an answer, as well as the answer itself.
- **Suggest** – is used where you are given an unseen resource in Papers 1 and 2, and asked to suggest what the reasons might be for something. You are being marked for your reasoning, on the basis of having studied Geography at A level.
- **Explain** – is used slightly differently in 6- and 8-mark questions.
  - With 6-mark questions, 'Explain' requires you to interpret a resource or use a resource to help stimulate your thinking.
  - With 8-mark questions, you are expected to know the answer without any stimulus material.
- **Evaluate** – is used for 20-mark questions in Papers 1 and 2, but for 18- and 24-mark questions in Paper 3. It's simply a device for asking questions differently. The meaning, and the degree of challenge, is still the same, and examiners would argue that it is the most challenging command word in any question.

Why should you have to understand assessment objectives?

An assessment objective (AO) is the key tool used by examiners to decide what you should know, should understand and be able to do after studying Geography A Level for two years. Read this section to get a clear idea of why it will help you to know and understand how you're being assessed, as well as the topics you have to learn.

What are the assessment objectives?

There are three assessment objectives in Geography A Level:

- **AO1** is about your **knowledge and understanding**. It might be your knowledge and understanding of places, processes or issues. It's basically the content of the two student books in this series!
- **AO2** is about the way in which you **interpret and apply** your understanding to situations. For example, you could be asked to consider an argument. Imagine a question such as 'Can child labour ever be justified?' You'd have to **weigh up evidence** (your knowledge and understanding: AO1) and frame it together into an **argument**. You might find there are points you could make that could support the use of child labour in certain circumstances, followed by others against its use. By the end, you might be able to **make a judgment** – perhaps in favour, perhaps against. That process – of using information to develop an argument – is what AO2 is all about. Most high-mark exam questions that you'll answer will have large numbers of marks allocated to AO2 – it's an essential A Level skill.

- **AO3** is about using **geographical skills** in formulating questions, in thinking about methods of data collection, in manipulating and presenting data, and in drawing conclusions. You should recognise that this is exactly what you've been doing in writing up your individual investigation, known as the Non-Examined Assessment (NEA). You will have to use those same skills in some situations in the exam as well.

- **AO1** questions use the command word 'Explain'. You also need knowledge and understanding for exam questions assessing AO2.
- **AO2** questions use the command words 'Suggest', 'Assess' or 'Evaluate', requiring you to use your knowledge and understanding (AO1).
- **AO3** questions use the command words 'Draw', 'Plot', 'Complete' or 'Calculate' (for statistics questions) or 'Analyse' (for questions about data interpretation).

	Objective	%	Marks
AO1	Demonstrate knowledge and understanding of places, environments, concepts, processes, interactions and change, at a variety of scales	34%	119
AO2	Apply knowledge and understanding in different contexts to interpret, analyse and evaluate geographical information and issues	40%	139
AO3	Use a variety of relevant quantitative, qualitative and fieldwork skills to: • investigate geographical questions and issues • interpret, analyse and evaluate data and evidence • construct arguments and draw conclusions	26%	92

▲ **Figure 7** Assessment objectives used in the exams

Marks allocated to assessment objectives

It helps to prepare for each exam if you know which AOs are being assessed (Figure 8 on page 10).

- For example, Papers 1 and 2 almost entirely assess **AO1** (knowledge and understanding) and **AO2** (application). This means that you really need to know your material, because AO1 counts for 46 of the total 105 marks, and AO2 for 55.
- Similarly, Figure 8 shows that Paper 3 (with the unseen Resource Booklet) has more **AO3** marks. That's because some questions will assess your ability to

read, manipulate and interpret data in the booklet. However, there are also a substantial number of **AO2** marks, because you'll be asked to analyse and make judgments about some of the issues, and **AO1** marks, because you'll be asked to relate what the issue, is about to other topics you've learned.

	AO1 marks	AO2 marks	AO3 marks	Total
Paper 1	46	55	4	105
Paper 2	46	55	4	105
Paper 3	19	21	30	70
NEA	8	8	54	70
Total	119	139	92	350

▲ **Figure 8** *Marks used for each assessment objective in the exams*

Command words, assessment objectives and marks

You know that some command words are more challenging than others. Figure 9 shows this in more detail – 'Explain' may carry up to 8 marks, while 'Evaluate' may carry 20.

However, several command words, shown in Figure 9, assess more than one assessment objective.

- For example, in Papers 1 and 2, you may be given a resource (e.g. Figure 10) and be asked to explain something about it for 6 marks.
- Imagine the question, *'With reference to Fig. 10, explain which geomorphic processes are the most influential in forming the landform shown.'*
- To do this, you need to recognise the landform and know something about its formation, so 3 of the 6 marks are

for AO1. The remaining marks are for applying what you know. For example, you might know several coastal processes but only some of these form arches. You therefore need to select the appropriate information – that's AO2.

As marks increase, so AO2 becomes more important.

- Figure 9 shows that 20-mark questions using 'Evaluate' are split – 5 marks for AO1 and 15 marks for AO2.
- A question such as *'Evaluate the extent to which urban regeneration depends upon rebranding for its success'* would therefore carry 5 marks for knowledge and understanding and 15 marks for AO2. (On page 15, you can see how examiners mark responses like this.)

Command word	Total mark	AO1	AO2	AO3
Explain	6	6		
Explain (using a resource)	6	3	3	
Explain	8	8		
Assess	12	3	9	
Papers 1 and 2 Evaluate	20	5	15	
Paper 3 – Analyse	8	4		4
Paper 3 – Evaluate	18	3	9	6
Paper 3 – Evaluate	24	4	12	8

◀ **Figure 9** *The balance of assessment objectives in questions using particular command words*

◀ **Figure 10** *Durdle Door in Dorset. This could be used as a resource for examiners to test your knowledge (AO1) and whether you can apply that knowledge to the photo (AO2)*

Revising examples and case studies

You probably remember learning several case studies for GCSE and it's possible you learnt a lot of detail off by heart. A Level is different. You need examples but not huge volumes of detail. It's much more important that you learn to argue a case based on concepts, with enough detail to support your case.

In the student book, Tectonic Hazards and Processes is one of the most popular A Level topics. Here are three stages to help you learn the essentials:

Examples you need to know for Tectonic Hazards and Processes

- One tectonic mega-disaster including its regional or global significance in terms of economic and human impacts (e.g. the 2004 Asian tsunami)
- One example of a multiple-hazard zone (e.g. the Philippines)

1 Build up a factfile

A basic hazard **factfile** should include:

- Location – Where is it? Can you locate it?
- What kind of hazard was it?
- When did it occur? Date and time?
- Was it a single event or one of several?
- Describe *briefly* what happened, e.g. Along which plate boundaries did an earthquake occur? What was its magnitude? Where was the epicentre?
- What caused it? What type of plate margin?

2 Know its impacts

The impacts of hazard events can be significant. What were the short-, medium- and long-term impacts? Be clear that you know what this means.

- **Short-term** – within the first month.
- **Medium-term** – within six months.
- **Long-term** – anything longer than six months.

Next, classify these into economic, social or environmental impacts.

- **Economic impacts** – e.g. jobs, businesses, trade, costs.
- **Social impacts** – people, health and housing.
- **Environmental impacts** – changes to the surrounding landscape.

You can now list these impacts and classify them, using a table like Figure 11. You need two impacts in each table 'cell'.

Impact	Immediate / short term	Medium term	Long term
Economic			
Social			
Environmental			

▲ *Figure 11* A table for classifying the impacts of a hazard event such as a volcanic eruption or an earthquake

3 Know the key ideas

Key ideas are particularly important because it's these that carry most marks. Tectonic Hazards and Processes is assessed by a 12-mark question using the command word 'Assess'. Study the examples of key ideas in the panel below and try to design your own exam questions. An example for Key idea 4 could be: *'Assess the role of governance in managing tectonic mega-disasters successfully.'*

For each one, you could then draw a mind map to show how you might answer the question.

Key ideas you need to understand for Tectonic Hazards and Processes

1 Theoretical frameworks, e.g. plate boundary theory
2 The relationship between hazards, vulnerability, resilience and disaster
3 Hazard profiles
4 The role of development and governance in understanding disaster impact, vulnerability and resilience
5 The trends and patterns of tectonic disasters
6 Theoretical frameworks for understanding hazards, e.g. Park's model
7 Hazard mitigation and adaptation strategies

Understanding key concepts

You'll spend most of your exam preparation time revising content for each topic. You'll also need to prepare for Paper 3 by revising details about players, attitudes and actions, and futures and uncertainties (see Chapter 6 in the *Geography for Edexcel A Level Year 2* Revision Guide).

However, you should also be aware of, and do some preparation to understand, the 14 key concepts (Figure 12) on which the whole A Level is based. All A Levels in England and Wales are based on these. Examiners could use the key concepts in any question and expect you to be able to say something about them.

Over to you

Copy and complete the table with examples in the right-hand column.

Concept	Definition	Example
Causality	Connections between cause and consequence as part of a process	
Equilibrium	A condition in which all influences acting within a system cancel each other or self-correct, so that systems are balanced	
Feedback	Responses to change; positive feedback causes further change and instability to a system, while negative feedback returns a system to equilibrium	
Globalisation	A process leading to greater international integration economically, culturally and demographically	
Identity	The beliefs, perceptions and characteristics that make one group of people or places seem different to others	
Inequality	Differences in opportunity, access to resources or outcomes (e.g. health) between different groups, at any scale	
Interdependence	Mutual reliance between groups; strongly linked to globalisation	
Mitigation and adaptation	Alternative approaches to management: preventative (mitigation) versus coping with (adaptation)	
Representation	How places or situations are portrayed to others, e.g. through stories, news items, photos, painting or other media	
Resilience	The ability to cope with change, e.g. resilience to global warming or to a hazard	
Risk	The potential or probability of harm / losing something of value	
Sustainability	The Brundtland definition refers to sustainability as *'development that meets the needs of the present without compromising the ability of future generations to meet their own needs'*, a contested term that can be interpreted in terms of economics, society, environment or politics	
Systems	Many interacting component parts, producing a complex 'whole', with inputs, flows, stores and outputs	
Threshold	A critical level in a system beyond which change is inevitable/irreversible	

▲ **Figure 12** *The 14 key concepts*

	Tectonic Processes and Hazards	Glaciated Landscapes and Change	Coastal Landscapes and Change	Globalisation	Regenerating Places	Diverse Places
Causality						
Equilibrium						
Feedback						
Globalisation						
Identity						
Inequality						
Interdependence						
Mitigation and adaptation						
Representation						
Resilience						
Risk						
Sustainability						
Systems						
Threshold						

▲ **Figure 13** *Which key concepts apply to the chapters in this Revision Guide*

Over to you

Copy and complete Figure 13 with ticks (√) to show which of the 14 key concepts apply to the six chapters in this Revision Guide.

However you look at it, revision can be dull! But, whatever revision you do should be **active**. This page will help you develop useful ways of revising.

Revising in groups

Working as a group is always better than alone. Try these ideas out.

Form a study group with friends Fix time with two or three friends to go through key topics. Do timed questions together, then mark them. Make lists of things you don't understand to ask your teacher.

Working together at home Message, Facetime or Skype friends and test each other. Go through questions together.

Test each other Make flash cards of key words and phrases, and construct mind maps of key concepts.

Know your key words Make lists of key words and phrases you need to know.

Revising, using the specification

Figure 14 shows an extract from the specification. Use it to make sure you revise everything that you should. The important features are:

Key ideas Examiners will use these to construct the longer 12- or 20-mark questions.

Detailed content Examiners will use this for the shorter questions.

Globe symbol Examiners can expect you to be able to use examples in questions set on the parts of the specification with a globe symbol.

Synoptic links These will be especially important in preparing for Paper 3.

This column contains the key ideas that will be assessed in longer exam questions, e.g. 'Assess the extent to which…'.

Broad enquiry questions form the basis of each sub-section of the specification.

This column contains all the key words and phrases you need to know. Most of the shorter 3-, 6- or 8-mark questions will come from this column.

Enquiry question 3: How successful is the management of tectonic hazards and disasters?	
Key idea	Detailed content
1.7 Understanding the complex trends and patterns for tectonic disasters helps explain differential impacts.	a. Tectonic disaster trends since 1960 (number of deaths, numbers affected, level of economic damage) in the context of overall disaster trends. (6); research into the accuracy and reliability of the data to interpret complex trends.
	b. Tectonic mega-disasters can have regional or even global significance in terms of economic and human impacts. (🌐 2014 Asian tsunami, 2010 Eyafjallajökull eruption in Iceland (global independence) and 2011 Japanese tsunami (energy policy))
	c. The concept of a multiple-hazard zone and how linked hydrometeorological hazards sometimes contribute to a tectonic disaster (🌐 the Philippines).
1.8 Theoretical frameworks can be used to understand the prediction, impact and management of tectonic hazards.	a. Prediction and forecasting (**P: role of scientists**) accuracy depend on the type and location of the tectonic hazard.
	b. The importance of different stages in the hazard management cycle (response, recovery, mitigation, preparedness). (**P: role of emergency planners**)
	c. Use of Park's Model to compare the response curve of hazard events, comparing areas at different stages of development.

The globe symbol means you should be taught examples, though any examples will do, not necessarily those shown.

The parts in bold are the synoptic links – P for players, A for attitudes and actions, F for futures and uncertainties.

▶ **Figure 14** How different parts of the specification are used by examiners

To gain the best marks possible in your exams, follow this advice.

1 Break down the question

Look at the example below (Figure 15). Try to break up questions like this. It will help you to focus on what the examiner is asking.

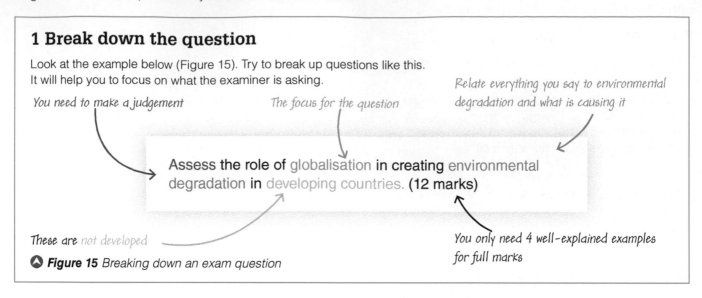

You need to make a judgement

The focus for the question

Relate everything you say to environmental degradation and what is causing it

Assess **the role of** globalisation **in creating** environmental degradation **in** developing countries. **(12 marks)**

These are not developed

You only need 4 well-explained examples for full marks

🔺 **Figure 15** *Breaking down an exam question*

2 Plan your answer

Planning your answer helps to organise your thoughts. Some people plan using a spider diagram, others just make a short list. Planning helps you to get the order of the answer right and makes sure you don't forget what to write. Figure 16 is an example of a plan.

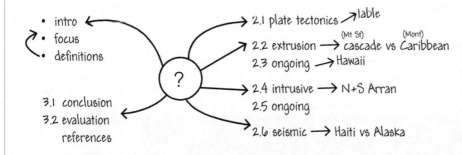

- intro
- focus
- definitions

3.1 conclusion
3.2 evaluation
references

?

→ 2.1 plate tectonics →lable
→ 2.2 extrusion → cascade vs Caribbean (Mt St) (Mont)
2.3 ongoing → Hawaii
→ 2.4 intrusive → N+S Arran
2.5 ongoing
→ 2.6 seismic → Haiti vs Alaska

◀ **Figure 16** *An example of a plan for the 12-mark question, 'Assess the extent to which theoretical frameworks help to understand the distribution of tectonic features.' You could add more details to this*

3 Get to know the mark scheme

Questions of 6 marks or more are assessed using level descriptors. These are the qualities that examiners look for in an answer. They are a series of criteria against which your answer will be judged. Between 6 and 12 marks, there are three levels – Level 3 is the highest. Above 12 marks, there are four levels, with Level 4 the highest.

As an example, the Level 4 descriptors for 20-mark questions, which use the command word 'Evaluate', include the following:

Demonstrates accurate and relevant geographical knowledge and understanding throughout – *i.e. you know your material well* (AO1)

Applies knowledge and understanding of geographical information/ideas to find logical and relevant connections/relationships – *i.e. you sequence your ideas so that geographical patterns are clear in your argument, rather than 'whatever comes next'* (AO2)

Applies knowledge and understanding of geographical information/ideas to produce a full and coherent interpretation that is supported by evidence – *i.e. your examples are planned into a conceptual argument and you use evidence selectively to prove a point rather than write down 'all I know about …'* (AO2)

Applies knowledge and understanding of geographical information/ideas to come to a rational, substantiated conclusion, fully supported by a balanced argument that is drawn together coherently – *i.e. you argue your points well and sustain your argument from start to finish* (AO2)

Top tips for exam success

You'll often hear students say 'Good luck' to each other as they enter the exam room. If you have done certain things, you won't need luck. Exam success comes from following a few rules. Students who perform well almost always follow the rules below.

They revise. Lack of revision always catches up with you. A levels demand knowledge and understanding of a great deal of content, so it's important to know your stuff!

They know which **topics** will be in each exam – for example, which exam tests physical or human geography, and which of their topics, such as tectonic processes and hazards, are on which paper.

They look at the **marks**, and know what sort of questions carry the highest marks. Similarly, they know and understand each **command word**.

They **practise** answers, often under timed conditions, for example, allowing 25 minutes for a 20-mark question.

They get **timing** right. Each paper is 2 hours and 15 minutes in length, but question length varies from one paper to another. For example, in Paper 3, you must take into account the time needed to read the Resource Booklet thoroughly. As a general rule on Papers 1 and 2, allow 13 minutes for every 10 marks.

They **answer everything that they should**, leaving no blanks. Even if unsure, they write something. Leaving a 12-mark answer blank could mean giving up a whole grade.

They write in **full sentences**. Single words or phrases are fine for 1–2 mark questions or for some of the quantitative skills questions. But 6-mark answers written in bullet points rarely score well and 20-mark 'Evaluate' answers even less so.

They learn **specific details** about case studies or examples. They take time to learn one or two statistics, names of places, and schemes. They don't just say 'in Africa'! Use specific place knowledge – you need this to earn the highest marks.

They get to know the **mark scheme** and the assessment objectives. Also get to know the **criteria** on which answers are judged and the differences between Levels 1, 2 and 3, or 4 where there is a Level 4 (see opposite).

Finally, they make sure they have a **timetable** that tells them exactly what time and which day each exam is on! Make sure you have one. Check and double-check it!

Chapter 1
Tectonic processes and hazards

What do you have to know?

This chapter studies primary tectonic hazards (earthquakes and volcanic eruptions), secondary hazards (e.g. tsunami) and the risks they pose.

The specification is framed around three enquiry questions:

1 Why are some locations more at risk from tectonic hazards?
2 Why do some tectonic hazards develop into disasters?
3 How successful is the management of tectonic hazards and disasters?

The table below should help you.

- Get to know the key ideas. They are important because 12-mark questions will be based on these.
- Copy the table and complete the key words and phrases by looking at Topic 1 in the specification. Section 1.1 has been done for you.

Key idea	Key words and phrases you need to know
1.1 The global distribution of tectonic hazards can be explained by plate boundary and other tectonic processes.	earthquakes; volcanic eruptions; tsunami; plate boundaries (divergent, convergent and conservative); oceanic and continental crust; causes of intra-plate earthquakes and volcanoes (hot spots, mantle plumes)
1.2 There are theoretical frameworks that attempt to explain plate movements.	
1.3 Physical processes explain the causes of tectonic hazards.	
1.4 Disaster occurrence can be explained by the relationship between hazards, vulnerability, resilience and disaster.	
1.5 Tectonic hazard profiles are important to an understanding of contrasting hazard impacts, vulnerability and resilience.	
1.6 Development and governance are important in understanding disaster impact and vulnerability and resilience.	
1.7 Understanding the complex trends and patterns for tectonic disasters helps explain differential impacts.	
1.8 Theoretical frameworks can be used to understand the predication, impact and management of tectonic hazards.	
1.9 Tectonic hazard impacts can be managed by a variety of mitigation and adaptation strategies, which vary in their effectiveness.	

You need to know:

- what is meant by the term 'natural hazards'
- why some hazards become disasters
- what makes countries like Nepal vulnerable.

Big idea

Natural hazards are different from disasters.

Hazards versus disasters

Nepal is located in southern Asia between India and China. An earthquake occurred on 25 April 2015, 80 km north-west of Kathmandu (see factfile).

- Nepal sits on a fault between the Indian and Eurasian **tectonic plates**. As the plates compress, pressure builds and releases.
- A naturally occurring event like the Nepal earthquake, with potential to cause loss of life or property, is called a **natural hazard**.
- Where a natural hazard causes social, environmental and economic damage, it becomes a **natural disaster** (i.e. when a vulnerable population can't cope using its own resources).
- People's ability to cope with hazard events is known as their **vulnerability**. The greater the hazard and more **vulnerable** people are, the greater the disaster.
- The overlap between the terms 'hazard' and 'disaster', and people's vulnerability, is shown on a model devised by Degg (Figure 1).

Factfile: Nepalese earthquake

Date: 25 April 2015
Magnitude: 7.8
Deaths: 8633
Injuries: Over 21 000
Homeless: 3 million people
Cost: Lost US$5 billion from its GDP (25% of the total)
Rebuilding cost: US$6.6 billion. Nepal will need to rely on foreign aid.

Interaction of humans and physical systems

Hazardous geophysical event: e.g. volcanic eruption, earthquake, tsunami

DISASTER

Vulnerable population: susceptible to human and/or economic loss because of where they live

▶ **Figure 1** Degg's disaster model showing how the overlap of a hazard and vulnerability can cause a disaster

Impacts of the Nepalese earthquake

The earthquake in Nepal was a major disaster.

- Nepal is extremely poor; half its population lives in poverty.
- The country's infrastructure (e.g. roads, water supplies) was severely damaged.
- Many buildings collapsed in the capital Kathmandu because they weren't built to withstand earthquakes.
- Over 100 aftershocks followed the initial earthquake, causing more destruction and deaths, and making rescue work dangerous.
- Nepal is mountainous. The shaking caused landslides, making rescue and aid efforts difficult in rural areas (Figure 2).
- Nepal's emergency services were unable to cope and relied on overseas countries and aid agencies.
- Tourism fell after the earthquake, putting people out of work.

▲ **Figure 2** The village of Barpak was one of many remote mountain villages destroyed by the earthquake

Ten-second summary

- There are differences between the terms 'natural hazard' and 'disaster'.
- Nepal is vulnerable to hazards.
- Hazard impacts may be economic, social or environmental.

Over to you

1 Close your book, then draw and label Degg's model.
2 Classify impacts of the Nepal earthquake as either economic, social or environmental. A Venn diagram will help.

You need to know:

- tectonic plate theory
- how the theory links to tectonic plate movement
- how different tectonic movements create different hazards.

Big idea

Tectonic plate theory helps to explain tectonic plate movement.

Earth structure and plate tectonics

The Earth has three layers (Figure 1):

- The **core** (the inner-most area) has two parts:
  - **Inner core** – solid centre, made mostly of iron, and the hottest part (about 6000°C).
  - **Outer core** – semi-molten, mostly liquid iron and nickel, with temperatures at 4500–6000°C.
- The **mantle** is the widest layer, surrounding the core. The upper mantle is solid; below it, the rock is semi-molten, forming the **asthenosphere**.
- The **crust** (the outer shell) has two types:
  - **Oceanic** – a thin dense layer (6–10 km thick), forming ocean floors.
  - **Continental** – a thicker, less dense layer (45–50 km), which makes up landmasses.

The crust and upper mantle make up the **lithosphere**, which is made up of seven major and several minor **tectonic plates** (large, irregularly-shaped slabs of crust). Study of these is called **plate tectonic theory**.

Plates move slowly over the asthenosphere, causing volcanic eruptions and earthquakes.

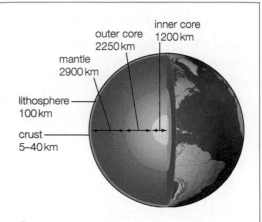

inner core
1200 km
outer core
2250 km
mantle
2900 km
lithosphere
100 km
crust
5–40 km

🔺 **Figure 1** *The components of the Earth's internal structure*

How tectonic plates move

Plate movement is driven by four processes.

Mantle convection

This is now much less accepted as a cause of plate movement. The theory is that heat produced by radioactive decay in the core heats the lower mantle, creating **convection currents** within the asthenosphere, causing plate movement.

North   North
N S N S N South   South N S N S N
Mid-Atlantic Ridge
Increasingly older basalts   Increasingly older basalts

🔺 **Figure 2** *Evidence for seafloor spreading. The age of the seafloor gets progressively older with distance from the mid-ocean ridge*

Seafloor spreading

Huge **mid-ocean ridges** form when hot **magma** is forced up from the asthenosphere and hardens, forming new oceanic crust. This **seafloor spreading** pushes tectonic plates apart (Figure 2).

Evidence comes from **palaeomagnetism**, a record of changes in the Earth's magnetic fields. When lava solidifies, minerals line up with the Earth's magnetic direction. Mid-ocean ridges show patterns of magnetic direction mirrored on each side of the ridges.

Subduction

As two plates move towards each other, one slides into the mantle in a **subduction zone**.

Slab pull

Slab pull is increasingly seen as driving plate movement. Newly formed oceanic material at mid-ocean ridges becomes denser and thicker as it cools, causing it to sink into the mantle, pulling the plate down.

Tectonic plate boundaries

Plate boundaries form where two tectonic plates meet. There are three boundary types:

- **Divergent** – two plates move apart.
- **Convergent** – two plates collide.
- **Conservative** – two plates slide past each other.

Each boundary creates distinct processes and landforms. Areas adjacent to plate boundaries are **plate margins**. There are four types: constructive, destructive, collision and transform.

Divergent boundaries (constructive margins)

Here, two plates diverge, forming new crust: mid-ocean ridges in oceans and rift valleys on continents.

- **Mid-ocean ridges** extend underwater as mountain chains with **transform faults** cutting across them. Mild shallow-focus earthquakes occur. Volcanic eruptions create **submarine volcanoes**, occasionally growing above sea level, e.g. Iceland.
- **Rift valleys** form on continents where the crust forms parallel cracks (**faults**) and the land between them collapses, creating steep valleys.

For a map of tectonic plate boundaries, see Figure 3 on page 9 of the student book. For diagrams explaining what happens at plant boundaries, see Figures 4–8 on pages 10–11 of the student book.

Convergent boundaries (destructive margins)

Here, plates converge and denser oceanic crust slides beneath lighter crust. There are three types:

- **Oceanic plate meets continental plate**
  Oceanic crust is denser than continental crust so it slides into the upper mantle and melts, marked by **deep ocean trenches**. Subduction leads to **fold mountains**, where plate collision causes folding. Constant movement creates continuous folding. Friction between plates causes major earthquakes in the **Benioff Zone**, where magma pushes through faults to reach the surface, forming explosive volcanic eruptions.
- **Oceanic plate meets oceanic plate**
  Subduction of the lighter of the two plates occurs. Deep ocean trenches and volcanoes form. Submarine volcanoes grow to form island volcanos (**island arcs**). Shallow- to deep-focus earthquakes occur. The 2004 Asian tsunami was caused in this way.
- **Two continental plates meet**
  A **collision margin** occurs, forcing rock up to form high fold mountains, e.g. Himalayas. There is no volcanic activity and earthquakes are shallow focus.

Conservative boundaries (transform margins)

When two plates slide past each other, they form a **conservative plate margin**, or **transform fault**. Although there is no volcanic activity, this movement creates powerful earthquakes. Plates sometimes stick, causing stresses to be released as shallow-focus earthquakes, e.g. Los Angles (1994) and San Francisco (1906, 1989).

 Ten-second summary

- Knowledge of the Earth's structure has led to tectonic plate theory.
- Plates move due to mantle convection, seafloor spreading, subduction and slab pull.
- Different tectonic plate boundaries (divergent, convergent and conservative) create different hazards.

 Over to you

Draw a composite labelled diagram to show the following: the Earth's inner and outer core, mantle, crust (both types), asthenosphere, lithosphere, subduction, slab pull, plate boundaries, mid-ocean ridges and rift valleys.

You need to know:

- the causes of earthquakes
- how they are measured
- their primary and secondary impacts.

Big idea

Physical processes explain the causes of earthquakes and their impacts.

Causes of earthquakes

On average, about 10 000 deaths are caused by earthquakes every year.

- 95% of earthquakes occur along plate boundaries. Friction between plates causes pressure to build, which is released as rocks fracture along **faults**, and energy is released as **seismic waves**.
- Pressure is released from the **focus** or **hypocentre**. The point on the Earth's surface directly above the focus is the **epicentre**, where most damage occurs.
- Seismic waves radiate from the focus. There are three main types (Figure 1). Primary (P) and secondary (S) waves are **body waves** as they travel through the Earth's body. Love (L) waves are **surface waves,** travelling along the Earth's surface.

📖 For a map of earthquakes and volcanoes around the world, see page 12 of the student book.

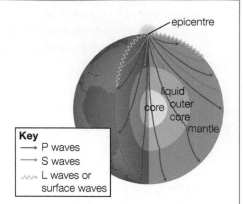

Key
⟶ P waves
⟶ S waves
〰 L waves or surface waves

🔺 **Figure 1** *The paths of P waves, S waves and L waves*

- **P waves** (primary or pressure waves) – move fastest and reach the surface first, travelling through solids and liquids, shaking backwards and forwards. They are the least damaging.
- **S waves** (secondary or shear waves) – move slower, travelling only through solids, moving with a sideways motion, shaking at right angles to the direction of travel. They are more damaging than P waves.
- **L waves** (surface love waves) – move slowest (last to arrive) but are the most damaging, shaking the ground from side to side. They are larger, focusing all their energy on the Earth's surface.

Seismic waves are measured using a **seismometer**. Results from several seismographs help to calculate the time, location and magnitude of an earthquake.

Measuring earthquakes

Earthquakes are measured by their **magnitude** and **intensity.**

- **Magnitude** is the energy released at the epicentre. Several scales measure it, but **Moment Magnitude Scale** (MMS) is better at measuring large earthquakes. It measures energy released by an earthquake, using information about:
  - seismic waves
  - rock movement
  - the fault surface broken by the earthquake
  - resistance of rocks affected.

The scale goes from 1 (smallest) to infinity, though generally stops at 10. The largest recorded earthquake was 9.5 in Chile in 1960. The scale is logarithmic (e.g. magnitude 5 is ten times more powerful than magnitude 4).

- **Earthquake intensity** is the impact of earthquakes on people, structures and the natural environment. The most common scale is the **Modified Mercalli Intensity Scale (MMIS)**, on a scale from I (hardly noticed) to XII (catastrophic).

Impacts of earthquakes

Impacts depend on: physical factors (e.g. magnitude, depth, distance from epicentre); human factors (e.g. level of economic development, response from emergency services); the impacts of indirect hazards (e.g. fires).

These impacts may be either **primary** or **secondary**, and also economic, social or environmental.

Primary impacts happen as a *direct* result of the earthquake (e.g. building or infrastructure collapse) and **crustal fracturing** caused by the energy released.

Secondary impacts result *from* the earthquake:

- **Liquefaction** occurs when surface rocks become more liquid than solid during movement. Buildings and roads sink, and power and gas lines break, causing fires.
- **Landslides** and **avalanches** are caused by the ground shaking. This causes a large portion of earthquake damage and injuries.
- **Tsunami** (large waves generated by earthquakes) cause coastal flooding.

Loma Prieta earthquake

- Date and time: 5:04 pm, 17 October 1989
- Magnitude: 6.9
- Location: Loma Prieta, near San Francisco
- Areas affected: Marina District was worst hit. Built on landfill, its soft, sandy soils liquefied, causing buildings and roads to collapse. The Cypress freeway (built on soft muds) collapsed (Figure 2), causing 42 of 67 earthquake-related deaths.

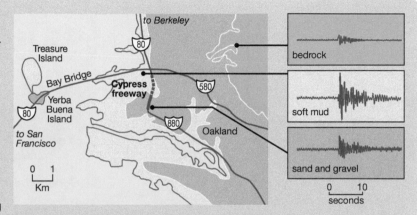

 Figure 2 *Earth movement around Cypress freeway during the 1989 earthquake, shown on seismograms*

Aftershocks

When the Earth readjusts along the faultline after a major earthquake, smaller earthquakes called **aftershocks** occur. The larger the initial earthquake, the larger and more numerous the aftershocks.

- Aftershocks cause additional damage (e.g. to building structures weakened by the initial earthquake), and hamper rescue efforts.
- The aftershock striking Christchurch, New Zealand in 2011 caused more damage and loss of life than the initial 2010 earthquake. It was shallower, causing more shaking, and was closer to the city.

Intra-plate earthquakes

Not all earthquakes happen at plate margins. **Intra-plate earthquakes** occur in the middle of plates. Their cause is uncertain, but is most likely stresses from ancient faults becoming active again. Their distribution is more random, making their prediction harder.

Earthquake prediction

There is no accurate method of prediction, but understanding plate tectonics helps to identify areas at risk.

- Most earthquakes occur along plate boundaries. Places that have had one big earthquake are likely to have another.
- **Forecasting** is general. The US Geological Survey forecasts a 67% chance of another 'serious' earthquake striking San Francisco by 2040.
- Research focuses on 'warning signs' (**precursors**), e.g. foreshocks (small shocks before larger ones), but none has proven reliable.

 Ten-second summary

- Earthquakes are caused by plate slippage, with seismic waves causing damage.
- Earthquake severity is measured using MMS and MMIS scales.
- Impacts are primary or secondary, and economic, social or environmental.

Over to you

Use one earthquake case study to categorise its impacts as:

a primary or secondary
b economic, social or environmental.

You need to know:

- the causes of volcanic eruptions
- how they are measured
- their primary and secondary impacts.

Big idea

Physical processes explain the causes of volcanic eruptions and their impacts.

Iceland: Eyjafjallajökull erupts

- Date: April 2010
- Location: Eyjafjallajökull, Iceland
- Impact: Because of atmospheric circulation, ash clouds from Eyjafjallajökull (Figure 1) affected flights over northern Europe for a week.
- 100 000 flights were cancelled, affecting 10 million people, losing airlines US$1.7 billion in revenues.
- Global airline capacity was cut by 30%, European capacity by 75%.
- Impacts of European flight cancellations were felt in Kenya, where 20% of the economy is based on exporting green vegetables and flowers to Europe, costing US$1.3 million daily in lost revenue.

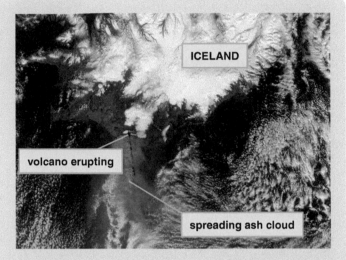

Figure 1 *Thick ash cloud from Eyjafjallajökull in 2010*

Causes of volcanic eruptions

Volcanoes are fractures in the Earth's crust through which lava, ash and gases erupt.

- Like earthquakes, they are associated with plate margins.
- As plates move, pressure builds and hot magma and gases rise from the upper mantle to the crust. When magma reaches the surface, it is called **lava**.
- As lava cools, it forms rock so, as volcanoes erupt, they grow.

More than 500 million people are at risk of hazards posed by volcanoes.

- Since 1800, 260 000 people have died from volcanic eruptions.
- 1900 volcanoes are now considered active, likely to erupt sometime.

Primary hazards

- **Lava flows** occur on the surface, reaching 1170°C, with many taking a long time to cool. Most are not a threat to people because they move slowly. However, some fast-flowing basaltic lava flows destroy everything in their path.
- **Pyroclastic flows** are a fast-moving, destructive mix of dense, hot (700°C) rock, ash and gases exploded from a volcano. They are very dangerous.
- **Ash and tephra falls** are volcanic rock fragments and ash blasted during eruptions, which can damage structures, injure or kill. Ash can travel thousands of kilometres, causing poor visibility and slippery roads.
- **Gas eruptions** are dissolved gases released during an eruption, including water vapour, carbon dioxide and sulphur dioxide. They can travel huge distances and be a major hazard.

Secondary hazards

- **Lahars** are flows of rock, mud and water down volcanic slopes, which can be hundreds of metres wide, flowing tens of metres per second. They form when eruptions melt snow and ice, or when heavy rainfall during an eruption erodes loose rock and soil.
- **Jökulhlaups** (glacial outburst floods) are formed when heat from an eruption melts snow and ice. They are sudden releases of water, rock, gravel and ice, and very dangerous.

Impacts of volcanic eruptions

Volcanic eruptions and secondary hazards can have enormous impacts on people, the economy and the environment, as Montserrat shows.

Montserrat: volcano erupts

The island is part of an arc in the Caribbean, formed where the Atlantic Plate subducts beneath the Caribbean Plate.

On 18 July 1995, the Soufriére Hills volcano began to erupt ash and dust. Eruptions continued for five years (the worst in 1997). Pyroclastic flows affected much of the island. Very little lava erupted. The volcano is still active.

Impacts included:

- Dozens died; two-thirds of its 11 000 people were evacuated, most permanently.
- The capital was destroyed (Figure 2).
- Two-thirds of houses were buried by ash or flattened.
- Three-quarters of infrastructure was destroyed.
- The tourist industry collapsed, causing unemployment.
- Farmland and rainforests were destroyed.
- Young people emigrated, leaving an ageing population.

Figure 2 *After Monserrat's capital Plymouth was covered in up to 3 metres of ash, it had to be abandoned*

Measuring volcanic eruptions

Volcanic eruptions vary, from gentle with small amounts of lava to huge explosions that eject enough ash to block sunlight.

Volcanologists use the **Volcanic Explosivity Index (VEI)** to measure the magnitude of an eruption on a logarithmic scale from 0 (non-explosive) to 8 (extremely large). The VEI scale is based on:

- volume of tephra and ash, pyroclastic flows, etc.
- height of material ejected into the atmosphere
- duration of the eruption
- observations (e.g. 'gentle', 'explosive').

Hot spots

Most volcanoes are located along plate margins, but **hot spots** are exceptions.

- Hot spots form in the middle of a plate, where rising **plumes** of magma then erupt on the ocean floor, forming a volcano.
- As a plate moves over a hot spot, the volcano is carried away with it, and a new volcano replaces it. A chain of volcanic islands is created, e.g. Hawaiian Islands.

Predicting volcanic eruptions

Eruptions can be predicted with some accuracy. Using seismometers, GPS and satellite-based radar, volcanologists can monitor volcanoes for signs of an eruption, e.g.

- small earthquakes – as rising magma causes small but detectable earthquakes
- the changing shape and tilt of the volcano – as upwards pressure of magma swells the volcano surface.

But predictions aren't 100% accurate and few volcanoes are monitored.

 Ten-second summary

- Volcanic eruptions are caused by plate movement.
- The severity of eruption is measured using the VEI.
- Impacts may be primary or secondary, and economic, social or environmental.

Over to you

Use one volcanic eruption case study to categorise its impacts as:

a primary or secondary
b economic, social or environmental.

You need to know:
- the causes of tsunami
- why they are difficult to predict
- their impacts.

Big idea
Physical processes explain the causes of tsunami and their impacts.

Causes of tsunami

A tsunami is a series of waves caused by undersea earthquakes along subduction zones, landslides or volcanic eruptions, or much less often by meteor or asteroid strikes.

- Most are three metres high or less, but large ones can reach 30 metres.
- They tend to occur along plate boundaries, particularly around the Pacific Basin's 'Ring of Fire'.

- If caused by an earthquake, the energy released causes the sea floor to lift, displacing the **water column** above, which then forms tsunami waves.
- A wave moves quickly (up to 800 km/h). As its crest nears the shore, it produces a vacuum effect, drawing water out to sea, exposing the sea floor.

For more detail on the formation of tsunami, see Figure 2 on page 20 of the student book.

Impacts of tsunami

Tsunami can be very destructive (Figure 1). The tall, fast-moving waves can destroy everything in their path.

- They can travel for several miles inland, sweeping away buildings, trees and bridges, as well as soil, and foundations of buildings and bridges.
- They change the landscape and small islands may be completely destroyed.
- Most tsunami-related deaths are from drowning but some are caused by collapsing buildings or trees. Flooding also causes contamination of food and water, leading to illness.

Figure 1 *A boat and other debris washed inland by the 2011 Japanese tsunami*

Predicting tsunami

The inability to predict earthquakes also means there is little way of predicting tsunami. However, tsunami **early warning systems** are now in place in the Pacific and Indian Oceans. These use seismic sensors to detect undersea earthquakes, as well as DART (Deep-ocean Assessment and Reporting of Tsunami) (Figure 2).

DART stations use sensors and buoys to monitor changes in sea level. As tsunami waves are detected, information is sent via satellite to warning centres, which use computer modelling to estimate tsunami size and direction, and inform areas at risk.

The Indian Ocean has had an early warning system since 2006. In 2004, when the Indian Ocean tsunami hit Sri Lanka before the warning system was in place, 31 000 people died.

Japan has the world's most extensive tsunami warning system. In 2011, Japan's Meteorological Agency issued a tsunami warning within three minutes of the earthquake. But it underestimated the size of the tsunami, so few took steps to prepare or evacuate.

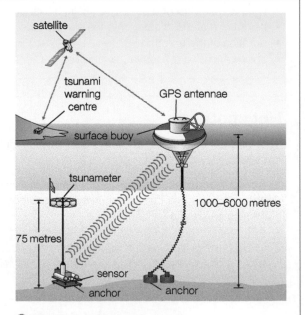

Figure 2 *The DART II system*

Indian Ocean tsunami, 26 December 2004

Background

- The earthquake causing the tsunami was magnitude 9.0–9.3. It caused one of the world's worst disasters.
- Upward thrust lifted the floor of the Indian Ocean by about 15 metres.
- The proximity of the epicentre to densely populated communities was significant. The earliest waves struck Banda Ache, Indonesia and were 17 metres high.
- The low-lying coastlines of neighbouring countries allowed the tsunami to travel inland.
- There was no early warning system.
- Many countries affected were LICs, without resources to spend on tsunami protection.
- Coastal mangroves had been destroyed to allow tourist development, reducing protection.

Affected country	Dead and missing
Indonesia	236 169
Sri Lanka	31 147
India	16 513
Thailand	5395
Somalia	150
The Maldives	82
Malaysia	68
Burma (Myanmar)	61
Tanzania	10
The Seychelles	3
Bangladesh	2
Kenya	1
Total	**289 601**

 Figure 3 *Distribution of dead and missing people in the 2004 Asian tsunami*

See Figure 6 on page 22 of the student book for areas affected by the tsunami.

Impacts

- 5 million people were affected in 14 countries.
- Over 230 000 people died, including 9000 tourists. 1.7 million were left homeless.
- Coastal settlements were devastated. 70% of people were killed. 1500 villages were destroyed in Sumatra.
- Infrastructure was destroyed.
- In Sri Lanka, more than 60% of the fishing fleet was destroyed. In Thailand, the tourism industry lost US$25 million per month and 120 000 people lost their jobs.
- Ecosystems such as mangroves and coral reefs were damaged.
- Water supplies and soil were contaminated by salt water.
- The total cost of damage exceeded US$10 billion.

BEFORE | AFTER

 Figure 4 *The settlement of Lhoknga, Sumatra before and after the tsunami*

Ten-second summary

- Tsunami are usually caused by plate movement.
- The severity of a tsunami varies with earthquake magnitude, distance from the origin and the level of development of the countries affected.
- Tsunami are difficult to predict.
- Impacts may be primary or secondary, and economic, social or environmental.

Over to you

a Copy and complete the table to show impacts of the 2004 Asian tsunami.

Impact	Short-term	Medium-term	Longer-term
Economic			
Social			
Environmental			

b Which impacts seem to have been the most serious? Explain why.

You need to know:

- the factors that determine vulnerability from, and resilience to, natural hazards and disasters.

Big idea

Vulnerability and resilience reflect how people are affected by natural hazard events.

Factors affecting risk from natural hazards

Risk from natural hazards, faced by people in different places, depends on several factors. It can be measured with the hazard risk formula (Figure 1). Based on this formula, it's possible to get an idea of the resilience of a place, i.e. its ability to protect lives, livelihoods and infrastructure from destruction and to recover after the event.

$$\text{Risk (R)} = \frac{\text{Hazard (H)} \times \text{Vulnerability (V)}}{\text{Capacity (C)}}$$

Figure 1 *The hazard risk formula assesses an area's risk from natural hazards*

- Some factors are physical, e.g. the magnitude, duration and time of day of the hazard (H in the formula).
- Human factors are also important in determining vulnerability (V) and the capacity to cope (C) (Figure 2).

Governance (local and national) and political conditions

- Existence and enforcement of **building codes and regulations**
- Quality of **infrastructure**, e.g. transport and power supplies
- Any disaster **preparedness plans**
- Efficiency of **emergency services** and response teams
- Quality of **communication systems**
- Existence of **public education and practice of hazard response**, e.g. earthquake drills
- Extent of **corruption** of businesses and government officials

Economic and social conditions

- Level of **wealth** affects people's ability to protect themselves and recover from a hazard
- **Access to education**
- Quality of **housing**
- Quality of **healthcare** systems
- **Income opportunities**

Physical and environmental conditions

- Density of **population**
- Speed of **urbanisation**
- Accessibility of an area

Figure 2 *A range of political, economic, social and physical conditions affecting hazard risk and the resilience of communities facing these risks*

 Ten-second summary

- Hazard risk can be expressed as the hazard risk formula.
- The vulnerability of any place from a natural hazard is affected by governance and political, economic, social, physical and environmental conditions.

 Over to you

1 Write out the hazard risk formula without looking at this page.
2 Draw three spider diagrams, one for each set of conditions shown in Figure 2. Show each factor as a separate 'leg', then annotate the diagram to show how each factor makes a community or country more or less vulnerable.

You need to know:

- the causes of earthquakes in Haiti, China and Japan
- about their impacts
- about responses to them
- how the Pressure Release Model (PAR) is used to assess a country's vulnerability.

Big idea

The impacts of and responses to earthquakes varies between developing, emerging and developed countries, and governments play a major role.

Earthquakes in contrasting countries

Japan (a developed economy), China (an emerging economy) and Haiti (an LIC) are located along tectonic plate margins, making them vulnerable to earthquakes. Each has experienced recent powerful earthquakes, with varying levels of destruction and loss (Figure 1). Of the three:

- Haiti's earthquake had the lowest magnitude, but the highest death toll.
- Japan's was 1000 times stronger than that affecting Haiti and included a tsunami.
- The impacts on Haiti were greater because it was more vulnerable.

	Haiti, 2010	Sichuan, China, 2008	Japan, 2011
Magnitude	7.0	7.9	9.0
Dead and missing	316 000 (govt. est) 220 000–250 000 (UN est)	87 150	19 848
Injuries	300 000	375 000	6 065
Homeless	1.3 million	5 million	130 927
Cost	US$14 billion	US$125.6 billion	US$240 billion
GDP per capita at time (2009 data)	US$1300	US$6600	US$39 473

 Figure 1 *Comparison of three earthquakes*

Haiti: a developing country

On 12 January 2010, a magnitude 7.0 earthquake struck Haiti (Figure 2). Deaths and injuries were among the worst on record because:

- The earthquake focus was shallow (13 km).
- Liquefaction caused building foundations to sink.
- The epicentre was only 24 km from the capital, Port-au-Prince.

However, the PAR model (Figure 3 on page 28) shows that the event became a disaster because:

- Haiti is poor, with resources spent on reducing poverty not earthquake preparation.
- High levels of government corruption reduced improvements to infrastructure or living standards.
- Buildings were poorly built because of lack of building regulations.
- Rescue teams found access difficult because of high population density.
- Few people knew what to do because they had not been prepared for disaster.

Much of Haiti's poor infrastructure was damaged.

- Its ports, main roads and only airport were damaged, preventing aid from arriving, increasing death tolls.
- 25% of government officials were killed and buildings destroyed, making recovery and relief difficult.
- A cholera outbreak lasted six years, affecting 720 000 people.

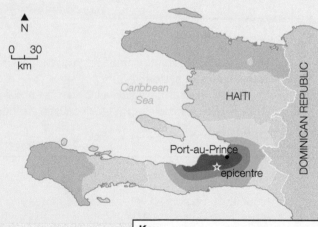 **Figure 2** *The earthquake's intensity and the number of Haitians at risk*

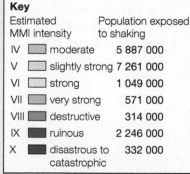

Key		
Estimated MMI intensity		Population exposed to shaking
IV	moderate	5 887 000
V	slightly strong	7 261 000
VI	strong	1 049 000
VII	very strong	571 000
VIII	destructive	314 000
IX	ruinous	2 246 000
X	disastrous to catastrophic	332 000

See Figure 3 on page 27 of the student book for the plate boundaries affecting Haiti.

Haiti's recovery

Haiti was still recovering in 2016. US$13 billion of aid was donated but most was controlled by international NGOs and governments. Haiti's government and organisations controlled less than 10%.

- With so many officials dead, NGOs were needed for emergency services.
- Concerns about corruption made many unwilling to give money to Haiti's government.
- Instead, they managed relief themselves, bringing staff from overseas. Many argue that this reduced Haiti's ability to recover or develop skills within its population.

Progress was slow, with 80 000 still in temporary housing in 2015. But there are improvements.

- New buildings, roads and schools have been built.
- Health statistics have improved.
- The government is stronger. In 2013, it responded to Hurricane Sandy, saving lives through warnings and organising aid afterwards.

Pressure and Release (PAR) Model

To protect people, it's important to understand a country's vulnerability to hazards. One tool that can be used is the **Pressure and Release (PAR)** model (Figure 3). It assumes that disasters happen when opposing forces interact. On one side, root causes, pressures and unsafe conditions create vulnerability; on the other is the hazard event itself.

- **Root causes** are political and economic systems that control who has power and resources.
- Through **dynamic pressures**, these lead to **unsafe conditions**, e.g. little is spent on enforcing building codes.
- The process is called the **progression of vulnerability**.

Root causes

- Haiti was heavily in debt. Some aid was spent in repayment of debts, not improving infrastructure.
- There was extensive corruption in Haiti's government.
- 80% of the population lived below the poverty line.
- 30–40% of the government budget was from foreign aid.

Dynamic pressures

- There was no building control, planning processes, disaster preparedness (including education) or management systems.
- There was rapid urbanisation with slum-like housing and high population density.
- Deforestation and soil degradation caused earthquake-related landslides.

Unsafe conditions

- Liquefaction amplified seismic waves, increasing the damage.
- Illegal housing was built in unsafe areas, e.g. on hillsides.
- Low GDP meant buildings were constructed cheaply and quickly.
- Poor infrastructure limited rescue efforts.
- Less than half the people had access to safe water or sanitation.

$$\text{Risk (R)} = \frac{\text{Hazard (H)} \times \text{Vulnerability (V)}}{\text{Capacity (C)}}$$

Hazards

Other climatic hazards (e.g. hurricanes) add to vulnerability from poverty and earthquake threats.

Figure 3 *The PAR model of Haiti's vulnerability*

China: an emerging country

On 12 May 2008, an earthquake of magnitude 7.9 struck Sichuan, a mountainous region in south-western China (Figure 4). Over 45.5 million people were affected. 5 million were made homeless (the highest number ever caused by a disaster). Landslides caused 25% of deaths.

As in Haiti:

- There were many injuries, and high economic cost.
- In some locations, corrupt government officials ignored building codes and poor construction led to building collapse.

However, the Sichuan earthquake was different.

- Its location meant most damage was rural, so there were fewer deaths.
- China is wealthier than Haiti, so able to pay for rescue efforts.
- China's strong central government responded with 130 000 soldiers and relief workers sent to affected areas.
- Medical services were restored, preventing disease.
- People at risk from landslides were relocated.
- The government gave US$10 billion for rebuilding.
- Within two weeks, temporary homes, roads and bridges were being built.

In general, China has tough building codes, with safer buildings, infrastructure and resources to respond to hazard events. The government rebuilt affected areas and within two years:

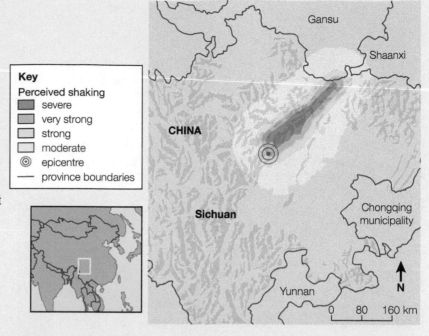

Key
Perceived shaking
- severe
- very strong
- strong
- moderate
- ◎ epicentre
- — province boundaries

🔺 **Figure 4** *The location and intensity of the 2008 Sichuan earthquake*

- 97% of reconstruction projects had started
- 99% of the 196 000 farmhouses destroyed had been rebuilt
- 216 transport projects were under construction or completed.

These made Sichuan more resilient to future hazard events.

However, the Chinese government was reluctant to accept overseas help, so several days lapsed before international rescue teams were allowed into the area.

Japan: a developed country

On 11 March 2011, a magnitude 9.0 earthquake struck under the Pacific, 100 km east of Sendai on Honshu's eastern coast (Figure 5).

- Seawater displacement caused 10-metre-high tsunami waves, moving at hundreds of kilometres per hour and surging 10 km inland.
- The Fukushima Daiichi Nuclear Power Plant was severely damaged, releasing dangerous levels of radiation and forcing the evacuation of 47 000 people. In 2018, a 20 km exclusion zone still existed.

Yet, despite this, Japan had fewer deaths and injuries than either Haiti or Sichuan.

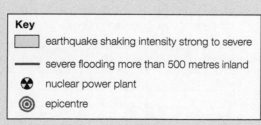

Key
- ▭ earthquake shaking intensity strong to severe
- — severe flooding more than 500 metres inland
- ☢ nuclear power plant
- ◎ epicentre

🔺 **Figure 5** *The epicentre of the earthquake and areas hit by tsunami waves*

Preparation

Japan has financial resources to prepare for hazard events, e.g.

- strict regulations, meaning buildings are better able to withstand earthquakes
- low levels of corruption, so building regulations are strictly enforced
- earthquake education, with drills practised in schools and businesses.

There are also well-developed disaster plans.

- 10-metre-high tsunami walls, evacuation shelters and marked evacuation routes helped to reduce deaths.
- Offices and homes have earthquake emergency kits of water and medical supplies.
- An early warning system detected the earthquake within minutes, giving people at least some warning.

Response

- 110 000 defence troops were immediately mobilised.
- All radio and TV stations broadcast what was happening and what to do.
- The Bank of Japan offered US$183 billion to keep the economy going.
- Japan accepted help from recovery teams from over 20 countries.

Impact on Japanese energy policy

Despite investment in disaster planning, Japan failed to consider the impact of a tsunami on the Fukushima nuclear power plant.

- The plant had not been built to withstand a large tsunami.
- A lack of safety procedures, preparation and oversight by the government contributed to events at Fukushima.

Before 2011, 27% of Japan's electricity came from nuclear energy. All 44 nuclear power stations were closed. By 2013, electricity from nuclear sources dropped to 1% and Japan imported fossil fuels (Figure 6). As a result:

- Electricity prices rose by 20%.
- Government debt rose.
- Greenhouse gas emissions increased.

By 2014, high electricity prices, a slowing economy and reliance on fossil fuels led the government to re-instate nuclear energy in its energy policy.

But the shutdown had a global impact: Germany closed all its nuclear plants.

2010

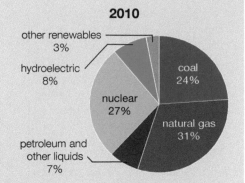

2013

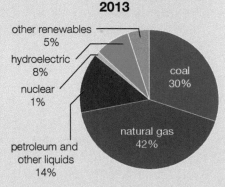

 Figure 6 *Japan's sources of electricity before and after the 2011 earthquake*

Over to you

1 Rewrite the PAR model (Figure 3) for:
  a Sichuan
  b Japan.

2 Copy and complete three tables like the one below to show the impacts of the hazard events in each of Haiti, Sichuan and Japan.

Impact	Short-term (a few weeks)	Medium-term (up to a few months)	Longer-term (up to a few years)
Economic			
Social			
Environmental			

Student Book
See pages 34–35

You need to know:
- about geophysical disaster trends
- what hazard profiles are.

Big idea
Tectonic hazard profiles help us to understand why hazard impacts and different disaster trends vary.

Geophysical disaster trends

Is the world becoming more hazardous? Since 1960, reported natural disasters have risen dramatically, although the number of **geophysical disasters** has remained fairly steady (Figure 1). This may be due to:

- better monitoring and reporting of disasters. In 2011, the tsunami following the Sendai earthquake was shown live on TV
- increased population density near rivers and coasts, leading to more awareness of storm or flood events.

Hazard trends sometimes appear complex because hazard events vary in their impacts, e.g.

- While some events cause large numbers of deaths, others of similar magnitude affect people in different ways (e.g. made them homeless) or cause greater economic damage.
- Hazard events that cause the greatest economic damage do not always cause the largest numbers of casualties.

HICs are better able to cope with hazard events.

- Fewer people are killed by disasters now because of early warning systems, better building structures and disaster management.
- Between 1994 and 2013, three times more people died per disaster in LICs than in HICs.

- Annual costs of disasters rose from US$20 billion in the 1990s to US$100 billion between 2000 and 2010 because of increased areas at risk.

However, disaster data are often incomplete or inaccurate because:

- The focus is on rescue not collecting data.
- There are differences in data collection, classification and defining 'damage'.
- Some hazards occur in remote areas.

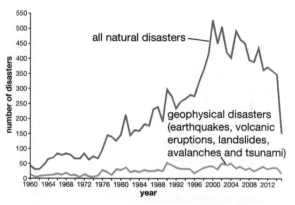

△ **Figure 1** *The numbers of reported natural disasters and reported geophysical disasters since 1960*

See Figures 2–4 on page 34 of the student book for hazard trends since 1960.

Hazard profiles

Hazard profiles show the characteristics of hazards and help to show ways that different events may be compared (Figure 2). They help governments and organisations to develop disaster plans.

- The profile for the 2004 Indian Ocean tsunami shows high magnitude, rapid onset and widespread extent, whereas the profile for Kilauea in Hawaii shows small magnitude and limited extent.

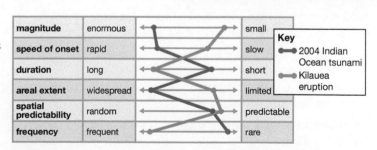

△ **Figure 2** *A hazard profile for the 2004 Indian Ocean tsunami and the ongoing eruption of Kilauea in Hawaii*

 Ten-second summary

- Better recording suggests there are more natural disasters.
- Impacts of hazards vary, e.g. the worst for deaths may not be the most costly.
- Hazard profiles help to compare different hazards.

Over to you

1 Explain how and why hazard impacts are:
   a similar
   b different.

2 Compare hazard profiles for any two hazards you have revised and explain the differences.

You need to know:

- what makes a multiple-hazard zone
- about one case study – the Philippines.

Big idea

Places facing more than one hazard are known as multiple-hazard zones.

Multiple natural hazards

Some countries are exposed to multiple natural hazards.

- They are known as **multiple-hazard zones** (or disaster hotspots). Hazards include **hydrometeorological** (e.g. flooding) as well as tectonic hazards.

- Identifying these helps decision-makers to understand and manage the hazards.
- Multiple-hazard zones need resources from international aid agencies to assist with disaster planning and prevention.

The Philippines: a multiple-hazard zone

Area: Consists of 7107 islands and is 25% bigger than the UK.
Population: 101 million (2015).
Wealth: GDP per capita PPP (2014) was US$7000; a lower middle-income country. 25% of the population lives in poverty.
Landscape: Mostly mountainous, with coastal lowlands.

The Philippines is one of the world's most disaster-prone countries (Figures 1 and 2). Hazards include:

- volcanoes (it has 47) and earthquakes, because of its location across a plate boundary on the 'Ring of Fire'
- tsunami – its coasts face the Pacific, the world's most tsunami-prone ocean
- typhoons – it sits within South-East Asia's major typhoon belts. It averages 15 typhoons annually due to its monsoon climate
- steep topography and deforestation make landslides common.

Vulnerability

A combination of rising population, urbanisation and poverty increases the Philippines's vulnerability.

- Economic development has led to rapid urbanisation and high population density.
- The poor live in coastal areas exposed to storm surges, flooding and tsunami.

One hazard event can have a knock-on effect. In 2006, an earthquake:

- killed 15, injured 100 and damaged or destroyed 800 buildings
- caused a 3 metre-high tsunami
- triggered landslides and created a flood.

Hazard	No. of events	Approx. Deaths	No. affected (millions)	Economic damage (US$)
Drought	8	8	6.5	64.5 million
Earthquake	23	9400	5.8	583 million
Flood	105	2100	21	3.4 billion
Landslide	32	2750	0.3	33 million
Volcanic eruption	22	1100	1.7	232 million
Tsunami	1	32	no data	no data
Storms / typhoons	345	45000	164	19.6 billion

Figure 1 *A summary of hazard events affecting the Philippines, 1960–2015*

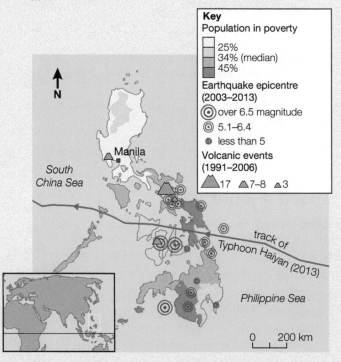

Figure 2 *The vulnerability of the Philippines to multiple natural hazards*

Ten-second summary

- Multiple-hazard zones are exposed to several hazard types.
- The Philippines suffers multiple hazards. Each impacts in its own right and also creates knock-on effects.

Over to you

Draw a detailed spider diagram to show ways in which the Philippines is a multiple-hazard zone.

You need to know:
- the stages of the hazard management cycle
- how Park's model (hazard response curve) is used.

Big idea
Theoretical models help to make
sense of managing hazards.

The hazard-management cycle

Hazard management is when governments and organisations work to protect people from natural hazards. The aim is to:

- minimise loss of life and property
- help those affected
- ensure rapid and effective recovery.

There are four stages (Figure 1).

- It's a circular process. Recovery (the final stage) should help to prevent future hazards (mitigation), e.g. making buildings earthquake-resistant.
- Each stage involves key players in planning, e.g. government, international organisations, businesses and community groups.

Stage	Focus	Actions	When?
Mitigation (Prevention) Preventing hazard events or minimising effects	• Identify potential hazards. • Try to reduce impact. • Aim to reduce loss of life and property.	• Plan zoning and land use. • Enforce building codes. • Build defences, e.g. tsunami walls.	Before and after hazard events
Preparedness Preparing to deal with an event	• Minimise loss of life and property. • Facilitate response and recovery phases.	• Be prepared, e.g. early warning systems. • Organise rescue services and evacuation. • Raise awareness, e.g. earthquake drills.	Before hazard events
Response Responding effectively to an event	• Aim to save lives, protect property, make affected areas safe and reduce losses.	• Search, rescue and evacuate where needed. • Restore infrastructure. • Ensure medical care and law enforcement continue.	During hazard events
Recovery Getting back to normal	**Short-term** • Focus on people's immediate needs. **Long-term** • Focus on people's needs, maybe for years. • Take steps to reduce vulnerability.	**Short-term** • Restore health, power, transport, water and financial services. • Remove debris and clean up. • Provide food and shelter. **Long-term** • Rebuild homes, schools, infrastructure and businesses.	After hazard events

 Figure 1 *Stages of the hazard management cycle*

See Figure 1 on page 38 of the student book for a diagram of the hazard-management cycle.

The Park model – the hazard-response curve

Park's hazard-response curve is a model that can be used to assess/compare how well places respond/recover after a hazard event (Figure 2). It shows that:

- The impact of an event changes, depending on magnitude, level of development and aid received.
- Response curves differ for different events. Some are sudden (e.g. 2010 Haiti earthquake) while others are slow (e.g. Montserrat's volcanic eruption).
- Wealthier countries recover more quickly.
- Where events affect several countries (e.g. 2004 Asian tsunami), each country has its own curve.

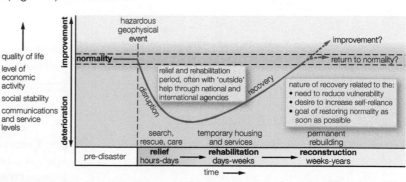 **Figure 2** *Park's hazard-response curve*

Ten-second summary

- The hazard-management cycle shows ways in which management can reduce hazard impacts.
- Park's model helps to assess recovery from a hazard event.

Over to you

Take any two hazard events.

a Assess each event in terms of how well the affected country was prepared (using the hazard-management cycle).
b Compare recoveries, using Park's model.

You need to know:
- about hazard mitigation strategies
- about hazard adaptation strategies
- the roles of different players involved.

Big idea

Different strategies are used to manage hazard events.

Hazard-mitigation strategies

These are strategies to avoid, delay or prevent hazard events.

- **Land-use zoning** involves local government **planners** regulating how land is used.
  - It is effective in protecting people and property in areas at risk from tectonic hazards, e.g. Mount Taranaki in New Zealand (Figure 1).
  - In areas of high risk from eruptions or tsunami, settlements and facilities such as nuclear power stations may be banned.

- **Diverting lava flows** is not usually successful as lava pathways are too unpredictable. However, barriers and channels to divert lava from Italy's Mount Etna were successful in 1983.

- **GIS mapping** can be used in all stages of disaster management, e.g. in deciding evacuation routes or rescue and recovery services. GIS information helped aid agencies in Nepal in 2015 to identify areas most affected by the earthquakes.

- **Hazard-resistant design and engineering defences** are researched by **engineers** to make buildings safer and prevent building collapse, which causes most deaths and damage from tectonic hazards.
  - New buildings can resist ground-shaking during earthquakes.
  - Roofs near volcanoes can be sloped to reduce ash build-up.
  - Buildings at risk from tsunami can be elevated.
  - Seawalls can be built to reduce tsunami impact.

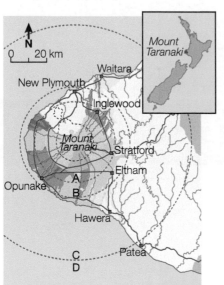

Key
Areas affected by future eruptions
1. likely to be affected most severely and most frequently by lava flows
☐ at Taranaki crater
☐ at Fanthams Peak crater

Possible depth of ash deposits
zone A 0.25m+
zone B 0.10–0.25m
zone C 0.01–0.10m
zone D < 0.01m

2. likely to be affected most severely and most frequently by debris, avalanches, lahars and floods
3. likely to be affected severely (moderate lahars and flood of intermediate frequency)
4. could be affected (unusually large but infrequent lahars and floods)
5. could be affected by lahars, volcanic debris or floods
6. unlikely to be affected by landslides, lahars or floods

◀ **Figure 1** *A volcanic hazard map for Mount Taranaki in New Zealand*

See Figures 2 and 3 on page 41 of the student book for illustrations of GIS mapping and earthquake-proof buildings

Hazard-adaptation strategies

These are strategies designed to reduce the impacts of hazard events.

- **Hi-tech monitoring** plays an important role in reducing vulnerability.
  - **GIS** creates hazard maps.
  - **Early-warning systems** detect signs of tectonic activity.
  - **Satellites** transmit data from monitoring equipment, e.g. the Pacific Tsunami Warning System (Figure 2).
  - **Mobile phones** communicate warnings and help co-ordinate preparation.

- **Crisis mapping** was used after Haiti's 2010 earthquake when members of Ushahidi, a free online resource, created **crowd-sourced**, interactive maps to provide information about people trapped or those in need of food and water. Volunteers plotted data and rescue workers used the maps in directing resources.
- **Modelling hazard impact** – using computer systems to model disasters and assess the impacts of a tsunami, for example.

- **Public education** reduces vulnerability and prevents hazards from becoming disasters through:
  - regular drills of procedures
  - encouraging homes and workplaces to create emergency kits
  - advising on buildings to withstand earthquakes.

Community-based preparation is an increasingly important part of this because local people often know best how to prepare, educate and identify vulnerable people.

See Figures 5 and 6 on page 43 of the student book for a computer model forecasting hazard impact and an example of community education.

▶ *Figure 2* *A tsunami early warning siren in Phuket, Thailand*

Key players in managing loss

Aid donors are essential.

- Aid has three stages in the hazard management cycle: **emergency** (e.g. food, clean water), **short-term** (e.g. shelter) or **longer-term** (e.g. reconstructing buildings).
- It is provided as cash, personnel, services and equipment.
- It is distributed to governments, which use it to manage recovery, or is controlled by aid agencies.
- It is provided by governments, IGOs (e.g. the UN) and NGOs.

NGOs play a crucial role where governments struggle or don't have resources (e.g. Haiti).

- They provide funds, co-ordinate search and rescue, and develop reconstruction plans.
- In Pakistan's 2005 earthquake (73 000 people died; roads, water and communication systems were destroyed), NGOs responded by providing tents, blankets, water, food and clothing.
- NGOs assist in recovery by building schools, medical centres and housing.

▲ *Figure 3* *Local people clear rubble in search of earthquake survivors in Afghanistan*

For **insurers**, the economic costs of natural disasters are staggering. US$54 billion was spent in 2011 as a result of earthquakes. Insurance cover helps recovery by providing individuals and businesses with money to repair and rebuild.

Local communities play critical roles in search and rescue, especially in remote or isolated communities (Figure 3).

 Ten-second summary

- Hazard mitigation includes land-use zoning, diverting lava flows, GIS mapping and hazard-resistant design.
- Hazard adaptation includes hi-tech monitoring, crisis mapping, modelling hazard impact and public education.
- Key players in managing loss include aid donors, NGOs, insurers and local communities.

 Over to you

1 Without looking, list:
   a four hazard-mitigation strategies
   b four hazard-adaptation strategies.

2 Construct a mind map to assess the strengths and weaknesses of these strategies.

Chapter 2
Glaciated landscapes and change

What do you have to know?

This chapter studies glaciated landscapes and the physical processes that form them. However, human activities are now changing these distinctive landscapes, which is threatening their future.

The specification is framed around four enquiry questions:

1. How has climate change influenced the formation of glaciated landscapes over time?
2. What processes operate within glacier systems?
3. How do glacial processes contribute to the formation of glacial landforms and landscapes?
4. How are glaciated landscapes used and managed today?

The table below should help you.

- Get to know the key ideas. They are important because 20-mark questions will be based on these.
- Copy the table and complete the key words and phrases by looking at Topic 2A in the specification. Section 2A.1 has been done for you.

Key idea	Key words and phrases you need to know
2A.1 The causes of longer and shorter climate change, which have led to icehouse-greenhouse changes.	glacial and inter-glacial periods, Pleistocene, climate change, Milankovitch cycles, solar output, volcanic eruptions, Loch Lomond Stadial (Pleistocene), Little Ice Age (Holocene)
2A.2 Present and past Pleistocene distribution of ice cover.	
2A.3 Periglacial processes produce distinctive landscapes.	
2A.4 Mass balance is important in understanding glacial dynamics and the operation of glaciers as systems.	
2A.5 Different processes explain glacial movement and variations in rates.	
2A.6 The glacier landform system.	
2A.7 Glacial erosion creates distinctive landforms and contributes to glaciated landscapes.	
2A.8 Glacial deposition creates distinctive landforms and contributes to glaciated landscapes.	
2A.9 Glacial meltwater plays a significant role in creating distinctive landforms and contributes to glaciated landscapes.	
2A.10 Glacial and periglacial landscapes have intrinsic cultural, economic and environmental value.	
2A.11 There are threats facing fragile active and relict glaciated upland landscapes.	
2A.12 Threats to glaciated landscapes can be managed using a spectrum of approaches.	

You need to know:

- that glaciated landscapes have environmental and cultural value
- that glaciated landscapes are important economically
- that glaciated landscapes face threats.

Big idea

- Glacial and periglacial landscapes have cultural, economic and environmental value.

Svalbard

The Svalbard islands lie in the Arctic Ocean, halfway between Norway and the North Pole.

- About 60% of Svalbard is covered by ice (Figure 1).
- Much of the rest of the land is bare ground.
- **Permafrost** exists almost everywhere.
- The islands contain the largest **fragile** wilderness area in Europe.
- Average temperatures range from 14°C in winter to 6°C in summer.

 Figure 1 *Much of Svalbard is covered by ice*

Threats to Svalbard

Coal mining

- Svalbard has valuable mineral reserves, but little extraction has occurred (apart from coal mining) due to climate and remoteness.
- The Norwegian state-owned mining company employs about a third of all workers on Svalbard to extract high-quality coal.
- The company faces difficulties, with job losses and calls from environmentalists to end mining.

Polar scientific research

- Svalbard has a history of polar scientific research.
- Much current research focuses on analysing atmospheric changes that might be linked to climate change.
- Research can result in environmental damage due to the infrastructure needed.

Tourism

- Opening the airport at Longyearbyen in 1975 caused tourist numbers to increase.
- Most visit Svalbard for the natural environment and wildlife, or to study the islands' history. Adventure tourism is increasing.
- Longyearbyen has seen a significant growth in tourist facilities.
- Tourists help the local economy but bring problems, e.g. oil spills and waste discharge from shipping.

Protecting Svalbard

Svalbard's economy depends on mining, research and tourism, but these activities threaten the fragile environment. The Svalbard Environmental Protection Act (2002) protects the natural environment and the islands' cultural heritage through national parks and nature reserves.

 Ten-second summary

- Svalbard is a glaciated landscape.
- Human activities threaten the wilderness on Svalbard.
- Two-thirds of Svalbard is protected through national parks and nature reserves.

 Over to you

Copy and complete the table below for Svalbard.

Economic activities	Why these are important	Threats posed

Student Book
See pages 50–53

You need to know:

- about the Pleistocene glaciation
- the long- and shorter-term factors leading to climate change
- the characteristics and causes of shorter-term climate events.

Big idea

A range of longer- and shorter-term factors lead to climate change.

The Pleistocene glaciation

The most recent period of ice activity occurred during the **Pleistocene** epoch of the **Quaternary** period (beginning approximately 2 million years ago). Temperatures on Earth fluctuated with cold periods (**glacials**) and warm periods (**interglacials**) (Figure 1).

- During glacial periods, ice masses spread south over Europe and North America (Figure 2).
- During warmer interglacial periods, much of the ice melted, and ice sheets and glaciers retreated.
- The southern-most regions of the UK stayed ice-free, despite being frozen and experiencing **periglacial** conditions (see Section 2.4).

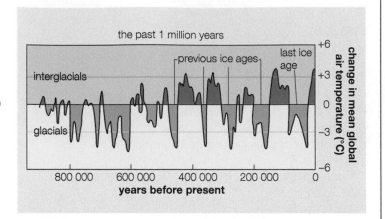

▲ **Figure 1** *General trends in mean global temperatures over the last 1 million years*

Pleistocene glaciation characteristics

- Temperatures fluctuated enough to allow a number of ice advances and retreats.
- The extent of the ice advance during each glacial was different.
- There were fluctuations within each major glacial. Pulses of ice advance are known as **stadials** and the warmer periods of retreat as **interstadials**.

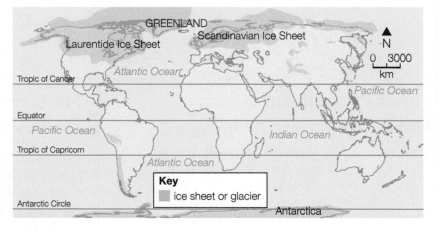

▲ **Figure 2** *The maximum extent of global ice coverage during the last glacial, about 20 000 years ago*

Long-term factors leading to climate change and glaciation

Milankovitch cycles

From 1912–41, Milankovitch (an astronomer and mathematician) carried out calculations showing that Earth's position in space, its tilt, and its orbit around the sun, all change. These changes, he claimed, affect the amount of incoming solar radiation – and where it falls on Earth's surface. They produce three main cycles (known as **Milankovitch cycles**), which could start, or end, an ice age.

📖 For details of Milankovitch cycles, see Figure 4 on page 51 of the student book.

Shorter-term factors

Variations in solar output

- Sunspots are 'flares' on the sun's surface indicating that the sun's radiation is more active than usual (Figure 3).
- High levels of sunspots increase emissions, raising Earth's average temperature.
- Changes caused by sunspots are small (+0.5 to – 0.5 °C globally).
- During longer-term cycles, sunspots seem to disappear almost completely.

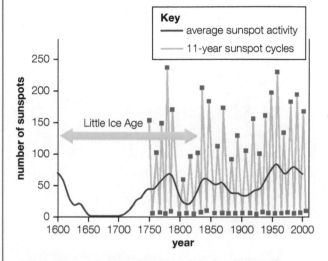

 Figure 3 *Sunspot cycles. Between 1650 and 1700, sunspots disappeared – this period is known as the Maunder Minimum*

Volcanic eruptions

- Big explosive volcanic eruptions can change Earth's climate.
- If ash and sulphur dioxide rise high enough, they will be spread around the stratosphere by high-level winds.
- Ash and gas reflect solar energy, preventing it from reaching Earth's surface.
- The temperature at Earth's surface falls and the planet will cool.

Shorter-term climate events

The Loch Lomond stadial

Between 26 000 and 10 000 years ago, climate fluctuations caused two stadials of ice advance, separated by an interstadial (Figure 4).

During the Loch Lomond stadial, it was sufficiently cold for an ice cap to develop over the uplands of western Scotland and for small tongues of ice to flow from the deeper corries of southern Scotland, the Lake District and Snowdonia.

Condition	Name	Years before the present
Ice retreating and disappearing	Flandrian interglacial	Since 10 000
Ice advance	Loch Lomond stadial	11 000–10 000
Ice retreat	Windermere interstadial	13 000–11 000
Ice advance	Dimlington stadial	26 000–13 000

 Figure 4 *The most recent glaciation in the British Isles*

The Little Ice Age

The Little Ice Age (during the **Holocene**) was a period of cooling that lasted from about 1550 to 1850. Suggested causes include volcanic activity and low levels of solar radiation (a lack of sunspot activity).

The Little Ice Age had a number of effects in Europe and North America.

- Winters were colder.
- Villages in the Swiss Alps were destroyed as glaciers advanced.
- Sea ice extended out from Iceland for miles.
- Greenland was cut off by ice from 1410 until the 1720s.
- European crop practices had to adapt to a shorter growing season.

🕐 **Ten-second summary**

- The most recent major period of ice activity occurred during the Pleistocene glaciation.
- There are a number of factors leading to climate change and glaciation.
- Shorter-term climate events include the Loch Lomond stadial and the Little Ice Age.

✏ **Over to you**

Write an explanation on the different factors leading to climate change and glaciation for a student beginning their A level course.

You need to know:
- the importance of the cryosphere and how ice masses are classified
- the distribution of cold environments
- about evidence for the Pleistocene glaciation.

Big idea
Glaciated landscapes provide evidence of past ice cover.

The cryosphere

The **cryosphere** is the frozen part of Earth's hydrological system. It is made up of:

- land surfaces, including ice sheets, ice caps, glaciers, areas of snow and permafrost
- frozen areas of oceans, lakes and rivers.

These areas act as **stores** within the global hydrological cycle.

Snow and ice reflect heat from the sun (the **albedo effect**), so the cryosphere helps to regulate temperatures on Earth.

Classifying ice masses

By scale and location

- **Ice sheets** – vast expanses of ice covering land surfaces.
- **Ice caps** – smaller masses of ice, often associated with mountain ranges.
- **Glaciers** – two main types: **cirque** or **corrie glaciers** (small) and **valley glaciers** (larger).
- **Ice fields** – areas of less than 50 000 km².

By thermal characteristics

- **Temperate glaciers** are **warm-based glaciers**. The base of the glacier is about the same temperature as the **pressure melting point**. They move 20–200 metres a year.
- Ice in **polar** or **cold-based glaciers** remains frozen at the base, so there is little movement.

Distribution of cold environments

There are four main types of cold environment and most are located in the far Northern Hemisphere.

- **Polar (high-latitude) regions** – areas of permanent ice (ice sheets of Antarctica and Greenland).
- **Periglacial (tundra) regions** – characterised by **permafrost** (found in Alaska, Scandinavia, etc.).
- **Alpine/mountain (high altitude) regions** – e.g. the European Alps, Himalayas. Glaciers and glaciated landscapes are found here.
- **Glacial environments** – at the edges of the ice sheets and in the highest mountainous regions, e.g. the Himalayas.

Evidence for the Pleistocene ice sheet

The UK's **relict glaciated landscapes** provide evidence that much of the country was covered by an ice sheet during the Pleistocene (Figure 1).

Erosional evidence	Found in the Cairngorms, Snowdonia, the Lake District. Includes: corries, arêtes, glacial troughs, plus roches moutonées, crag and tail, knock and lochan landscapes.
Depositional evidence	Drumlins (e.g. in the Vale of Eden, Cumbria), erratics (e.g. the Bowder Stone, the Lake District), moraine (the Cairngorms).
Meltwater evidence	Meltwater channels (e.g. Newtondale, North Yorkshire), glacial till (e.g. Holderness coast), eskers (e.g. Blakeney, Norfolk).

⌂ **Figure 1** *Evidence for the Pleistocene ice sheet in the UK*

See Sections 2.8, 2.9 and 2.10 for further details about erosional, depositional and meltwater landforms.

Ten-second summary

- The cryosphere is the frozen part of Earth's hydrological cycle. It helps to regulate temperatures on Earth.
- Ice masses can be classified by scale, location or thermal characteristics.
- Most of the world's cold environments are located at high latitudes.
- Relict glaciated landscapes in the UK provide evidence for the Pleistocene ice sheet.

Over to you

Explain the difference between:

a an ice sheet and a glacier
b valley glaciers and ice fields
c temperate and polar glaciers
d polar, periglacial and alpine regions.

You need to know:

- the distribution of past and present periglacial landscapes
- about periglacial processes
- periglacial landforms and landscapes are formed.

Big idea

Periglacial processes produce distinctive landscapes.

Where are periglacial landscapes found?

They are found on the fringes of polar glacial environments.

- They experience extreme cold for much of the year, with penetrating frosts and periodic snow cover, and are underlain by **permafrost**.
- Areas affected by permafrost today cover about 25% of the world's land area. In the past, they were more extensive.

Permafrost is where a layer of soil, sediment or rock below the ground surface remains almost permanently frozen. It reaches depths of 400–500 metres and may be continuous, discontinuous or sporadic, depending on the temperature. Cycles of freezing and thawing result in a saturated seasonal **active layer** in the summer.

For more on the past and present distribution of periglacial landscapes, and on permafrost, see pages 55 and 56 of the student book.

What are periglacial landscapes like?

- They are typically featureless plains, either strewn with rocks or covered in tundra vegetation (low-growing plants, e.g. mosses, lichens, dwarf shrubs).
- Soils are thin and water-logged.
- **Thaw lakes** form in summer when lying snow and the thin active layer melts.

Figure 1 *A periglacial landscape in the Northwest Territories of Canada*

Periglacial processes and landforms

For more detail on periglacial processes and landforms, see pages 58–60 of the student book.

In periglacial landscapes, most **geomorphological processes** are associated with frost, ice, snow and meltwater. The processes create a range of distinctive landforms (Figure 2, and Figure 3 on page 42).

Figure 2 *Periglacial landforms*

	Processes	Landforms
Frost	**Frost-shattering/freeze-thaw weathering** occurs as temperature falls to 0°C. Water freezes and expands by about 9%, exerting stress within rocks. Repeated freezing/thawing causes rock to break up.	**Talus/scree, blockfields, felsenmeer** (large areas of angular rock fragments)
Snow	**Nivation** processes are most active around the edges of snow patches.	**Nivation hollows** (if occupied by ice, they may enlarge to form corries)
Ground ice	Meltwater fills cracks, and freezes. Extensive ice wedges form polygonal patterns on the ground. Edges of polygons can be marked by stones due to **frost heave** (ice can form within pores or as ice needles in soil and sediment, which can force soil particles / stones upwards to the surface).	**Ice wedges, patterned ground, stones polygons** and **stripes**
Ground ice	Freezing water pushes overlying sediments upwards into a dome-shaped feature.	**Pingos:** • **open system** (or East Greenland type) • **closed system** (or Mackenzie type)
Meltwater	**Solifluction** is the slow downhill movement of saturated soil.	**Solifluction lobes**
Meltwater	**Meltwater erosion**	**Braided rivers** (flow across glacial outwash plains)
Wind	**Strong winds** pick up fine, loose material and deposit it as **loess**.	**Loess** covers large areas in the Mississippi-Missouri valley (USA) and north-west China.

🔺 **Figure 3** *Periglacial processes and landforms*

Baffin Island, Canada

Canada's Baffin Island has a typical periglacial landscape, with continuous permafrost and a tundra ecosystem (Figure 4).

- It has a number of ice caps.
- The topography ranges from rugged mountains to flat lowlands. The coastal strip is littered with lakes and ponds fed by run-off streams.
- Winters are long, dark and cold. During the short summer, the snow and soil layers above the permafrost (the active layer) melt. This creates and feeds a network of lakes, streams, rivers and wetlands. The waterlogged soil and 24-hour sunshine during summer boost rapid plant growth.

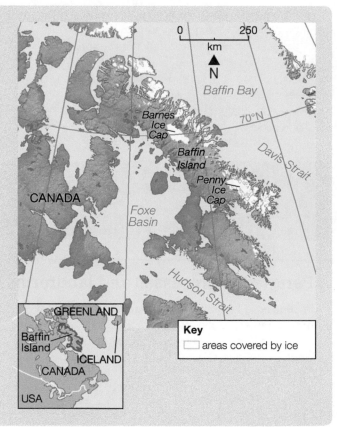

▶ **Figure 4** *Baffin Island, Canada*

 Ten-second summary

- Periglacial landscapes exist on the fringes of polar or glacial environments.
- Geomorphological processes in periglacial landscapes are mostly associated with frost, ice, snow and meltwater, and create a range of landforms.
- Periglacial landscapes are usually featureless plains with tundra vegetation. Baffin island is typical.

 Over to you

Create a spider diagram of the processes found in periglacial landscapes. Add further legs for the landforms created by those processes. Add as much detail as you can.

You need to know:

- about glacial mass balance and about glaciers as systems
- about accumulation and ablation (including their variations and impact on mass balance).

Big idea

Mass balance is important in understanding glacial dynamics and how glaciers operate as systems.

What is the glacial system?

A glacier is an **open system** with inputs from, and outputs to, other systems (Figure 1).

Inputs

- Direct **precipitation** (snowfall)
- **Avalanches**
- **Wind deposition**

Outputs

- **Water** (due to melting close to the glacier's snout)
- **Calving** – if the ice front extends over water, chunks may break off to form icebergs.
- **Evaporation** and **sublimation**

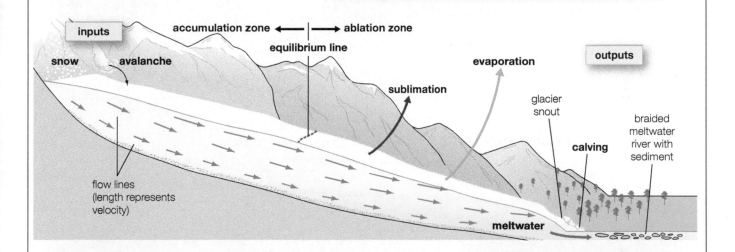

Energy

A glacier's mass combines with gravity to generate potential energy, which enables it to carry out erosion, transportation and deposition.

Stores/components

The main stores are snow and ice.

Flows/transfers

These include processes such as evaporation, sublimation, meltwater flow and glacial movement (**internal deformation** and **basal slip**) (see Section 2.6).

Feedback loops

Positive and **negative feedback loops** occur in glacial systems.

- Negative feedbacks regulate the system to establish balance and equilibrium.
- Positive feedbacks enhance and speed up processes, promoting rapid change.

Dynamic equilibrium

In a glacial system, the **equilibrium line** marks the boundary between the **accumulation zone** (glacial inputs) and the **ablation zone** (glacial outputs). If the glacier is in a state of balance (inputs equal outputs), the equilibrium line will remain in the same place. As this balance shifts, the equilibrium line will move up or down the glacier (known as **dynamic equilibrium**).

🜢 *Figure 1* The glacial system

Glacial mass balance

The balance between accumulation and ablation determines whether a glacier grows or shrinks, and is known as the **mass balance**. It is calculated by dividing the glacier into:

- The **accumulation zone** – there is a net *gain* of ice over the course of a year (inputs exceed outputs).
- The **ablation zone** – there is a net *loss* of ice during a year (outputs exceed inputs).

Mass balance varies over a year (Figure 2). Ablation is highest in summer; accumulation is highest in winter.

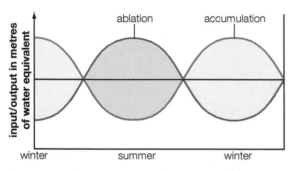

🔺 **Figure 2** *Seasonal variations in mass balance*

Why do glaciers advance and retreat?

They do so in response to long-term trends in mass balance.

- If *accumulation* exceeds *ablation* (a positive mass balance), the glacier's mass increases and it advances.
- If *ablation* exceeds *accumulation* (a negative mass balance), the glacier's mass decreases and it retreats.

Changes in a glacier's mass balance affect how it works and creates landforms.

- If the climate cools, the ice thickens, moves faster and advances. It erodes and transports more vigorously.
- During warmer periods a glacier shrinks and the ice becomes thinner and retreats. Its movement slows and it erodes and transports less debris, and deposits more.
- Changes in mass balance can occur year on year (with variations in ablation and accumulation), as well as over decades and much longer periods.

Glacier health: the Gulkana Glacier, Alaska

A glacier's health can be assessed by determining its mass balance.

The USGS has studied the Gulkana Glacier since the 1960s in order to measure changes in its mass balance (Figure 3), understand glacier dynamics and hydrology, and to assess the glacier's response to climate change.

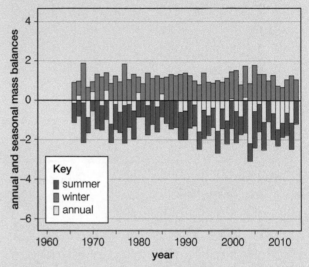

🔺 **Figure 3** *Changes in the Gulkana Glacier mass balance (1966–2014) in metres of water equivalent*

 Ten-second summary

- A glacier is an open system with positive and negative feedbacks.
- Glacial systems move towards a state of dynamic equilibrium.
- The change in the glacial ice budget is known as the mass balance.
- Glaciers advance and retreat in response to long-term trends in mass balance.
- A glacier's health can be assessed by determining its mass balance.

Over to you

Design your own diagram of the glacial system. Show the equilibrium line, accumulation zone and ablation zone. Add labels to show inputs, outputs, stores and flows.

You need to know:

- how glaciers move
- about factors that affect the rate of movement
- that polar and temperate glaciers move at different rates.

Big idea

Glaciers move as a result of different processes, and at different rates.

How does ice move?

Ice moves in two main ways: internal deformation and basal slip.

Internal deformation

This occurs through:

- *Inter-granular movement* – individual ice crystals slip and slide over each other.
- *Intra-granular movement* – individual ice crystals become deformed or fractured due to stresses within the ice. Gradually, the mass of ice moves downhill in response to gravity.

Basal slip

Basal slip (or sliding) usually occurs in a series of short jerks. This happens in temperate glaciers (where meltwater helps to lubricate the base of the ice).

- When a glacier encounters an obstacle, the increase in stress and pressure may result in **pressure melting**, allowing the glacier to move over the obstacle.
- The meltwater may refreeze on the downslope side.
- Melting and freezing that depends on pressure is called **regelation**. The associated movement is called **regelation creep**.
- Downhill movement can raise the temperature of the base ice (due to increased pressure and friction). This **positive feedback** may lead to further melting, allowing the glacier to slide more easily.

How fast do glaciers move?

- Polar glaciers move only a few metres a year and almost exclusively by internal deformation.
- Temperate glaciers move 2–200 metres a year due to internal deformation and basal slip.

Variations in the rate of movement

- Increases in gradient cause ice to flow faster. It becomes 'stretched' and thinner, and is known as **extensional flow**.
- Reductions in gradient force the glacier to slow down, pile up and thicken. This is called **compressional flow**.
- Between the zones of extensional and compressional flow, ice moves in a **rotational** way.

Compressional flow increases the mass and erosional power of the glacier. This can lead to a steeper gradient, faster extensional flow, a thinning of the ice and reduction in potential erosion – a **negative feedback** loop.

Factors affecting the rate of movement

- **Altitude**
- **Gravity** and **gradient** (**slope**)
- **Friction**
- **Ice mass**
- **Underlying surface permeability**
- **Meltwater**
- **Ice temperature**

Ten-second summary

- Ice moves due to internal deformation, basal slip and regelation creep.
- Ice flows in different ways: extensional flow, compressional flow and in a rotational way.
- Positive and negative feedbacks can affect glacier movement.
- The rate of ice movement is affected by many factors.
- Polar and temperate glaciers move at different rates.

Over to you

Draw a spider diagram to show the factors determining:

a how glaciers move
b the rate at which they move.

Student Book
See pages 68–69

You need to know:
- how glacial processes alter landscapes
- that glacial landforms develop at different scales
- how glacial landforms can be used to study the extent of ice cover.

Big idea
Glacial processes produce landforms in various environments at different scales.

Glacial processes

Erosion and entrainment

There are two main types of glacial erosion:

- **Abrasion** occurs because of **entrainment** (ice includes angular frost-shattered material, which scours the landscape).
- **Plucking** or **quarrying**.

For details on abrasion and plucking, see Section 2.8 (and opposite).

Transportation

Glaciers transport material in three ways:

- **Supraglacial** – mainly weathered material carried on top of the ice.
- **Englacial** – material carried within the ice.
- **Subglacial** – material carried below the ice.

Water also transports material on top of glaciers and, by flowing down crevasses or holes (moulins), it transports material into and beneath the ice. Meltwater streams may carry material under the ice.

Deposition

Sediment is deposited when ice melts, mainly in the ablation zone close to the glacier's snout. Water can then carry the sediment further away.

Glacial landforms

For a diagram of glacial landforms, see Figure 3 on page 69 of the student book.

Glacial landforms develop in different environments, where different processes operate (Figure 1).

Environment	Processes and landforms	Scale
Subglacial – below a glacier or ice-sheet.	• Erosion forms **striations** (scratches) and roche moutonée. • Meltwater can create eskers due to deposition.	Localised or **micro-scales**, or can extend over **meso-scales**
Marginal – at the sides or end of a glacier or ice sheet.	• Weathering and deposition create landforms such as moraines.	**Micro-** or **meso-scale**
Proglacial – in front of, at or immediately beyond the margin of a glacier or ice sheet.	• Fluvioglacial processes create outwash plains, meltwater channels and proglacial lakes.	**Meso-scale**
Periglacial – at the edges of ice sheets and glaciers.	• Processes are mostly associated with frost, ice and snow, plus meltwater. • Landforms include blockfields, ice wedges and pingos.	**Macro-scale**

🔺 **Figure 1** *Landforms, environments, processes and scale*

Landform evidence

Landforms created by glacial processes differ in upland and lowland areas.

- Upland areas are characterised by erosional landforms, e.g. corries, glacial troughs and hanging valleys.
- Lowland areas tend to have more depositional landforms, e.g. outwash plains and glacial till deposits.

Landforms such as glacial and meltwater deposits tell us where glaciers operated in the British Isles and how far they extended.

See Figure 4 on page 69 of the student book for a map of glacial and meltwater deposits across the British Isles.

Ten-second summary

- Glaciers alter landscapes by the processes of erosion, entrainment, transport and deposition.
- Glacial landforms develop in different environments and at different scales.
- Glacial landforms can provide evidence of where glaciers operated and how far they extended.

Over to you

Close your book. Name:

a two types of glacial erosion
b three ways that glaciers transport material
c four named environments where glacial processes operate.

You need to know:

- how glacial erosional processes lead to the formation of landforms associated with glaciers
- how landforms are caused by ice sheet scouring.

Big idea

Glacial erosion creates distinctive landforms, which together create glaciated landscapes.

Glacial erosion

- **Abrasion** is the sandpapering effect of ice as it grinds over and scours a landscape. **Freeze-thaw** weathering creates sharp rock fragments, which, if trapped under the ice, become abrasive tools. Large rocks carried beneath the ice can scratch the bedrock to form **striations**.

- **Plucking** or **quarrying** occurs when basal meltwater freezes around part of the underlying bedrock. Loose rock is 'plucked' away as the glacier moves.

  The weight and pressure of ice can also **crush** the surface of bedrock.

How glacially eroded landforms develop

Different erosional landforms are associated with **cirque** and **valley glaciers** (Figure 1).

- **Cirque (or corrie) glaciers** are small masses of ice that occupy armchair-shaped hollows in mountains. They often feed valley glaciers.

- **Valley glaciers** are larger masses of ice that move down from either an ice field or a cirque. They usually follow former river courses.

Figure 3 (on page 48) shows a typical glacially eroded landscape.

Landform	Formation
Corries	<ul><li>A **corrie** is an enlarged hollow on a mountainside. It has a steep, cliff-like back wall, often with a large pile of scree at its base. There is usually a raised rock lip at the front of the hollow, which acts as a dam to trap water and form a small lake (**tarn**) (Figure 2).</li><li>A range of processes including nivation, abrasion, plucking and rotational sliding help corries develop.</li></ul> ⬤ **Figure 2** *A corrie in post-glacial conditions*
Arêtes and pyramidal peaks	<ul><li>If two neighbouring glaciers/corries cut back into a mountainside, they leave a narrow, knife-edge ridge called an **arête**.</li><li>If three or more corries erode back-to-back, the ridge becomes a **pyramidal peak**.</li></ul>
Glacial troughs	<ul><li>The vast mass of a glacier enables it to erode steep-sided, mainly flat-bottomed, deep **glacial troughs**. They tend to be straight, because of the power and inflexibility of the glaciers that erode them. The main process of erosion is abrasion through basal slip.</li><li>Glacial troughs may contain deep, narrow lakes called **ribbon lakes**. These result from enhanced vertical erosion.</li></ul>
Hanging valleys	<ul><li>When the ice of a small glacier in a tributary valley melts, the smaller valley is left 'hanging' above the main valley.</li></ul>
Truncated spurs	<ul><li>When an upland river valley is occupied by ice, the rigid glacier cuts off or 'truncates' the tips of interlocking spurs as it moves downhill, leaving behind steep cliffs called **truncated spurs**.</li><li>Abrasion is the main process in the formation of truncated spurs, but plucking also takes place on bare rocky surfaces. Sub-aerial processes, e.g. freeze-thaw and rockfalls (mass movement), also act on the exposed rocky cliffs above the ice.</li></ul>

⬤ **Figure 1** *The formation of glacially eroded landforms*

For more detail on corrie formation and orientation, see page 71 of the student book.

Red Tarn – a corrie lake

Corrie – a scooped out hollow in the landscape

Striding Edge – a classic example of an arête

Glacial troughs or U-shaped valleys.

⬢ **Figure 3** *The Lake District – a glacially eroded landscape*

The formation of landforms due to ice-sheet scouring

Landform	Formation
Crag and tail	Formed when a large resistant object, or crag, obstructs the flow of a glacier. Ice is forced around the obstruction, eroding weaker rock. Material in the lee of the obstruction is protected by the crag, leading to the formation of a sloping **tail** of deposited material.
Roche moutonnée	These are outcrops of rock on the valley floor that were sculpted by moving ice (Figure 4). glacial ice plucking / direction of ice flow / abrasion plucked boulders bedrock ⬢ **Figure 4** *How abrasion and plucking form a roche moutonée*
Knock and lochan landscapes	This is a glacially scoured lowland area, which has alternating roches moutonnées (known as cnoc or knock) and eroded hollows, which often contain small lakes (lochans). They are most often found where alternate resistant and weakly jointed rocks allow differential erosion.

⬢ **Figure 5** *The formation of landforms due to ice-sheet scouring*

 Ten-second summary

- There are two main types of glacial erosion – abrasion and plucking (quarrying) – which lead to the formation of erosional landforms.
- Other glacial erosional landforms are created as a result of ice-sheet scouring.

 Over to you

Create a revision sticky note for each of the landforms included in this section. Stick them on your bedroom wall and test yourself daily.

You need to know:

- how glacial depositional features are formed
- how glacial landforms can be used to work out ice extent and movement.

Big idea

Glacial deposition creates distinctive landforms and contributes to glaciated landscapes.

Glacial deposition landscapes

Glaciers transport rock debris from upland erosion and deposit it on valley floors or lowland plains to create two types of depositional features :

- ice-contact depositional features
- lowland depositional features.

Many glaciers are now shrinking and retreating, revealing some classic landforms (Figure 1).

▶ **Figure 1** *The Steingletscher Glacier in the Swiss Alps retreats to reveal classic landforms of glacial deposition*

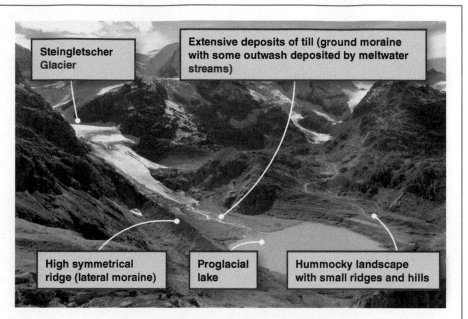

Steingletscher Glacier

Extensive deposits of till (ground moraine with some outwash deposited by meltwater streams)

High symmetrical ridge (lateral moraine)

Proglacial lake

Hummocky landscape with small ridges and hills

Ice-contact depositional features

Moraines

Moraine is a generic term for landforms associated with the deposition of **till** from within, on top of, and below, a glacier. It consists of poorly sorted, mainly angular sediments. Figure 2 shows the characteristics of moraines and other depositional features.

There are several types of moraine: **ground, terminal, recessional, lateral, medial**.

Drumlins

A typical **drumlin** is an oval or 'egg-shaped' hill, made up of glacial till and aligned in the direction of ice flow.

- Drumlins vary in size, but are commonly 30–50 metres high and 500–1000 metres long.
- They usually occur in clusters or 'swarms' on flat valley floors or lowland plains in previously glaciated regions.
- Some have a rocky core (with sediment moulded around it), but most don't.
- Some consist partly of fluvial sediments as well as glacial till.

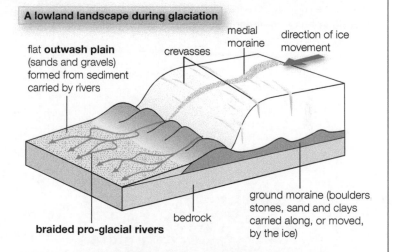

A lowland landscape during glaciation

flat **outwash plain** (sands and gravels) formed from sediment carried by rivers

crevasses

medial moraine

direction of ice movement

ground moraine (boulders, stones, sand and clays carried along, or moved, by the ice)

bedrock

braided pro-glacial rivers

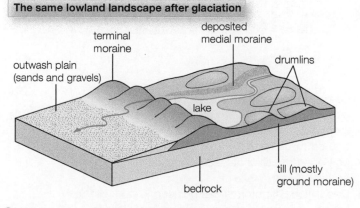

The same lowland landscape after glaciation

terminal moraine

outwash plain (sands and gravels)

deposited medial moraine

drumlins

lake

till (mostly ground moraine)

bedrock

▲ **Figure 2** *Landforms of glacial deposition*

Lowland depositional features

Till plains

A **till plain** is created by the melting of a large ice sheet that detached from a glacier. Meltwater streams can carry sediment for many kilometres, eventually depositing it as a vast, gravelly, well-sorted **outwash plain**. Till levels out the landscape, making it mostly flat.

Till is characteristically angular and poorly sorted, and be divided into:

- **lodgement till** (dropped by moving glaciers)
- **ablation till** (dropped by stagnant or retreating ice).

Erratics

An **erratic** is a boulder or rock fragment deposited far from its origin. They provide clues to help work out the direction of previous ice movement. By working out the source of the erratic, the origin of the ice can be established.

Working out previous ice extent, movement and origins

It's possible to use glacial deposition landforms to work out how far ice extended and in which direction it flowed, as well as the origin of depositional features. For example:

- Terminal moraine provides evidence of a glacier's furthest extent.
- A crag and tail indicates the direction of ice flow.

Drumlin orientation

Drumlins cover much of the Vale of Eden in Cumbria. Debris-laden ice moved east through the area. The drumlin field it left behind (Figure 3) indicates the direction of movement and the volume of material eroded, transported and deposited.

See Figures 6 and 8 on pages 76 and 77 of the student book for a map of the movement of ice in the Vale of Eden and an OS map extract of the area shown in Figure 3.

For details on till fabric analysis, which can be used to suggest the direction of ice flow, see page 75 of the student book.

Using GIS

Geographers at the University of Sheffield have produced a glacial GIS map of the UK, showing the extent of ice during the last Ice Age. The GIS contains over 20 000 features split into thematic layers such as moraines, eskers, drumlins, meltwater channels and ice-dammed lakes.

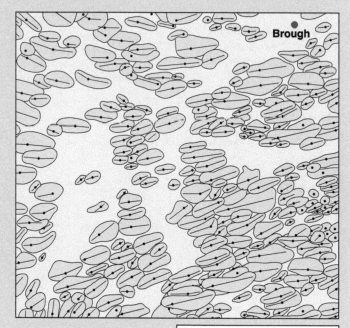

 Figure 3 *The drumlin field near Brough, Cumbria*

Key

drumlin outline with crestline and highest point

Ten-second summary

- Glaciers transport rock debris from areas of upland erosion and deposit it on valley floors or lowland plains.
- Ice contact depositional features include moraines and drumlins.
- Lowland depositional features include till plains and erratics.
- Depositional landforms can be used to work out how far ice extended and in which direction it flowed.

Over to you

Draw an annotated diagram to describe the typical location and formation of different types of moraine. Refer to pages 74–75 of the student book if necessary.

You need to know:

- how water moves within the glacial system
- how fluvio-glacial features form
- the different characteristics of glacial and fluvio-glacial deposits.

Big idea

Glacial meltwater plays a significant role in creating distinctive landforms and landscapes.

What are fluvio-glacial landscapes and processes?

Landscapes created by fluvio- (river) glacial processes are associated with flowing water (meltwater) in glacial or periglacial environments.

- Meltwater erodes, transports and deposits sediment.

- Ice can produce huge quantities of water when it melts. This can flow on top of a glacier (in **supraglacial** channels), within the ice (**englacial** channels) or under it (**sub-glacial** channels).

There are two distinct types of fluvio-glacial landforms:

- ice-contact features
- pro-glacial features.

Ice-contact features

Eskers

Eskers are long, winding ridges of sand and gravel, up to 30 metres high and several kilometres long. They run roughly parallel to the valley sides, which suggests that they were formed by sub-glacial river deposition, as the glacier retreated.

Kames

Kames form on the ice surface. They consist largely of sand and gravel deposited by streams in the final stages of a glacial period. There are three different types (Figure 1):

- **Kame terraces** result from the infilling of a marginal glacial lake.
- **Kame deltas** form when a stream deposits material on entering a marginal lake.
- **Crevasse kames** form as a result of sediment deposited in surface crevasses.

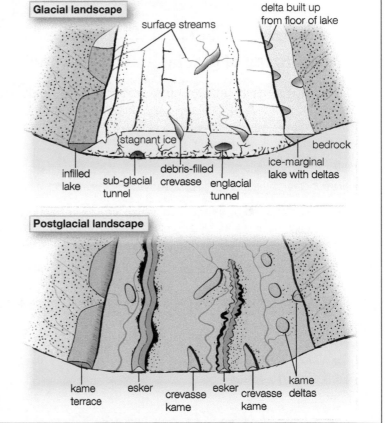

▶ *Figure 1 How kames and eskers are formed*

Pro-glacial features

Outwash plains (sandur)

An **outwash plain** is an extensive, gently sloping area of sands and gravels that forms in front of a glacier. It results from the 'outwash' of material carried by meltwater streams and rivers. At the end of a glacial period, huge quantities of material are spread over the outwash plain by torrents of meltwater.

Kettle holes

These form when large blocks of ice (left behind when glaciers retreat) are covered by deposits from meltwater streams. When the ice melts, a depression is left, which fills with water, forming a kettle hole.

Meltwater channels and pro-glacial lakes

When ice sheets expand and dam rivers, they can create **pro-glacial lakes**. During the last ice advance, the North Yorkshire Moors remained largely unglaciated, forming an ice-free island (Figure 2).

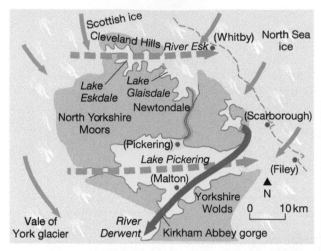

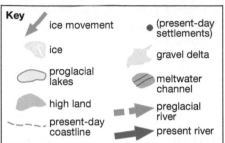

- Lake Eskdale (a pro-glacial lake) formed when the North Sea ice sheet blocked the mouth of the River Esk. The level of the lake rose and the overflow river flowed through Lake Glaisdale before cutting the deep, narrow, steep-sided, flat-floored Newtondale Valley – a **meltwater channel**.

 Figure 2 Pro-glacial lakes and meltwater channels in North Yorkshire

Characteristics of glacial and fluvio-glacial deposits

Glacial deposits	Fluvio-glacial deposits
Unstratified; difficult to identify layers	**Stratified**; there are layers due to seasonal variations in sediment accumulation
Unsorted; melting ice deposits material regardless of size	**Sorted**; as meltwater loses energy, larger rocks and boulders are deposited first
Material is angular (from physical weathering and erosion), and various shapes and sizes	Material is smooth and rounded (due to attrition), sorted and **graded**

Figure 3 Characteristics of glacial and fluvio-glacial deposits

Deposits can be affected by **imbrication**, where they are orientated (aligned), overlapping each other like toppled dominoes. They can be aligned in two ways with the longest axis either parallel or perpendicular to the direction of meltwater flow.

Fluvio-glacial landforms have different depositional characteristics (Figure 4).

See student book page 81 for more on imbrication, and Figure 8 to see how fieldwork can measure sediment size and shape change with distance across an outwash plain.

Landform	Characteristics
Eskers	Consist of sorted coarse material, usually sand and gravel; often stratified
Kames	Mounds of coarse sand and gravel; sorted and stratified
Outwash plains	Material is sorted; largest material is deposited first (near the glacier); the finest material (clay) travels furthest across the plain
Meltwater streams	Those crossing outwash plains are **braided**. Variations in volume of meltwater lead to streams becoming choked with coarse material

Figure 4 Fluvio-glacial landform characteristics

 Ten-second summary

- Landscapes created by fluvio-glacial processes are associated with flowing water in glacial and periglacial environments.
- Water moves within the glacial system as supraglacial, englacial and subglacial flows.
- There are two types of fluvio-glacial landforms – ice-contact features and proglacial features.
- Glacial and fluvio-glacial deposits and landforms have different characteristics.

Over to you

1 Draw a spider diagram to show the features of a fluvio-glacial landscape.
2 Describe in detail how two of those features are formed. Learn this!

You need to know:

- that relict glaciated landscapes have environmental and economic value
- the challenges and threats glaciated landscapes face from human activities
- that different players (stakeholders) are involved in managing the challenges.

Big idea

Glaciated landscapes have an intrinsic value but face threats, which need to be managed.

Glaciated landscapes in the UK

A number of distinctive **relict glaciated landscapes** still exist in the UK. These developed when the climate was much colder and a large part of the UK was covered by ice (Figure 1). They consist of upland and low-lying areas. The Lake District is an example of a relict glaciated landscape.

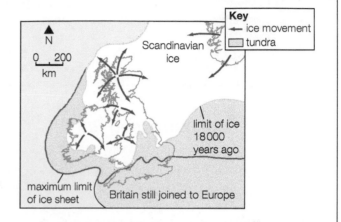

▶ **Figure 1** *The ice cover over the UK 18 000 years ago created the relict glaciated landscapes of today*

The Lake District – opportunities and threats

Tourism

Over 16 million people visit the Lake District every year (Figure 2).

- The local economy benefits (by £1.1 billion in 2014), helping to support shops, hotels, pubs, etc.
- Many services that benefit tourists also benefit local people (e.g. better public transport).
- Some of the money that tourists bring in is used to protect the environment.
- Tourism provides over 16 000 jobs in the National Park and boosts the local economy through the **multiplier effect**.

Both the landscape and ecology of the Lake District are fragile and under threat from overuse. Activities such as walking and climbing can lead to footpath erosion, trampling and littering, challenging the area's **resilience**. There are other problems:

- congestion and pollution
- poorly paid and seasonal jobs
- rising house prices.

Farming

The trend in the Lake District is towards fewer and larger hill farms, but there has also been an increase in the number of the smallest holdings.

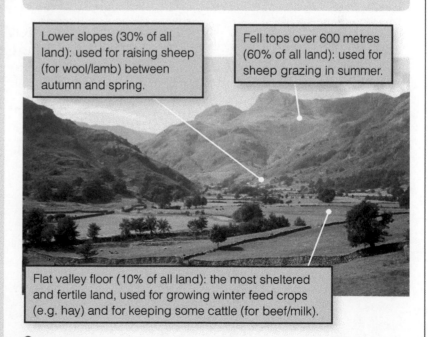

Lower slopes (30% of all land): used for raising sheep (for wool/lamb) between autumn and spring.

Fell tops over 600 metres (60% of all land): used for sheep grazing in summer.

Flat valley floor (10% of all land): the most sheltered and fertile land, used for growing winter feed crops (e.g. hay) and for keeping some cattle (for beef/milk).

▲ **Figure 2** *Farming in Langdale, typical of the landscape that tourists come to see*

Water storage and forestry

A dam was built at the northern end of Thirlmere in the late 19th century to provide water for Manchester. This glacial valley originally contained two small tarns and a hamlet. Creating the reservoir submerged these. The aqueduct built to carry water to Manchester now also carries water from Haweswater.

The land surrounding Thirlmere is forested. This reduces soil erosion and also generates income from selling the timber.

Key players in managing for the future

The Lake District National Park Partnership was formed in 2006 to give organisations involved in the Park more say in its management. Twenty-five organisations (**stakeholders)** are involved, representing the public, private, community and voluntary sectors.

The future

The Lake District National Park Partnership's vision for 2030 includes strategies covering everything from biodiversity, water quality, farming, skills and training to employment, housing, transport and tourism.

Ten-second summary

- The Lake District is a relict glaciated landscape that is important economically and environmentally.
- The landscape and ecosystem of the Lake District are under threat.
- The Lake District National Park Partnership works with a range of stakeholders to manage the Lake District for the future.

Over to you

In no more than six sentences or bullet points, summarise:

a the threats the Lake District faces
b how the threats can be managed.

The Lake District and climate change

Climate change threatens the Lake District's unique landscape and fragile ecosystem. Some of the likely impacts are:

- the loss of indigenous plant and animal species, and an increase in non-native species
- the gradual movement of habitats from lower to higher altitudes
- heavy rain will wash soil and farm chemicals into the lakes
- peat on the fells will dry out in warmer summers (releasing stored carbon) and dry moorland and forests will be more prone to fires.

Tackling climate change

In 2008, the 'Low-carbon Lake District' initiative was launched. It's designed to tackle climate change and is working with local businesses, communities and other agencies to reduce greenhouse gases (**mitigation**) and prepare for the impacts of climate change (**adaptation**). Work to create a low-carbon Lake District (Figure 3) includes initiatives such as reducing emissions and transforming how visitors get to, and around, the Lake District.

◀ *Figure 3*
Low-carbon Lake District official logo

You need to know:

- that active glaciated landscapes have environmental, cultural and economic value
- the threats glaciated landscapes face from human activities and natural hazards
- the different stakeholders involved in managing the threats and challenges.

Big idea

Glaciated landscapes have an intrinsic value but face threats that need to be managed.

The Sagarmatha National Park

The Sagarmatha National Park (Figure 1) includes Mount Everest. It's an active glaciated area with a fragile ecosystem. It is a destination for mountain tourism and is home to over 6000 Sherpa people. It has enormous environmental and cultural value, but faces a range of threats.

▶ **Figure 1** *The Sagarmatha National Park in Nepal, an area of dramatic mountains, glaciers and deep valleys*

Tourism – opportunities and threats

The tourism industry in the Sagarmatha National Park is largely in the hands of the Sherpa people. Tourism has boosted the local economy, leading to improved standards of living, better healthcare, education and infrastructure. But it causes environmental damage and socio-economic change, including:

- footpath erosion and the construction of illegal trails
- water pollution
- problems with waste disposal
- increased demand for forest products and new hotels and lodges
- changes to the Sherpa way of life.

Increasing numbers of people are turning to adventure tourism in wilderness areas such as the Himalayas. The chances of a successful 'summit' of Everest have increased with improvements in equipment, the number, and standard, of mountain guides and weather forecasting. But increasing numbers of climbers on Everest are causing problems, and some are calling for tighter controls.

For more information on the Sherpa people, see page 86 of the student book.

Climate change and shrinking glaciers

In 2013, researchers found that some glaciers around Mount Everest had shrunk by 13% in the last 50 years and that glaciers may be disappearing faster each year. The consequences include:

- impacts on farming and hydropower generation downstream
- the creation of lakes dammed by glacial debris, which if breached can cause catastrophic floods.

Avalanches

On 25 April 2015, a 7.8 magnitude earthquake struck Nepal. Although Mount Everest was approximately 220 km east of the epicentre, the earthquake triggered several large avalanches on and around the mountain, killing at least 22 people.

Deforestation and landslides

Less than 30% of Nepal's forest remains. Deforestation has been caused by:

- farming
- the use of firewood as fuel
- clearing forested areas to build roads, reservoirs, etc.

The consequences include:

- loss of wildlife habitats and biodiversity
- infertile soil (nutrients are washed away from exposed soil)
- erosion of exposed soil
- increased risk of landslides
- disruption of the water cycle.

Glacial outburst floods

When glaciers melt, glacial mass balances (see Section 2.5) are disrupted, which risks disrupting the hydrological cycle (see Section 2.13). In the long term, this will lead to a reduced water supply downstream.

- Lakes blocked by moraine are often found close to the snouts of valley glaciers. These lakes grow if glaciers melt and the moraine wall may collapse, causing a **glacial lake outburst flood** (Figure 2).
- Increases in summer warming could lead to floods and the possible destruction of settlements, irrigation systems and water supplies downstream.

Key

The potential triggers for glacial lake outburst floods include:

A glacier calving – when chunks of ice falls off a glacier

B icefall from hanging glaciers

C rock, ice or snow avalanches

D moraine degradation

E rapid input of water from glacial streams

F seismic activity

 Figure 2 *Potential triggers of glacial lake outburst floods*

Managing Sagarmatha for the future

Sustainable management is vital in order to balance the exploitation and preservation of the National Park. Some of the **stakeholders** involved are shown below. Strategies used to protect the Park include:

- re-establishing forests
- banning goats (to protect vegetation)
- increasing the use of kerosene for cooking and heating
- building micro hydroelectric power stations to generate electricity.

The key players

The stakeholders include:

- **Global organisations** – it's a UNESCO World Heritage Site.
- **Government agencies** – the National Park was established in 1976.
- **Local residents and stakeholders** include local residents (mostly Sherpa), incomers who have set up small businesses and a Park Advisory Committee.
- **NGOs**, e.g. the Sagarmatha Pollution Control Committee.

 Ten-second summary

- Everest and the Sagarmatha National Park is an active glaciated area.
- It faces threats from human activities and natural hazards.
- Sustainable management is vital in order to balance exploitation and preservation of the National Park.

Over to you

Complete a table with three columns to show:

a the value of the Sagarmatha National Park landscape

b the threats it faces

c strategies for managing it.

You need to know:

- that tundra ecosystems play an important role in maintaining natural systems
- about the impact of global warming on glaciated areas and tundra ecosystems
- how legal frameworks are used to protect landscapes.

Big idea

Glacial and periglacial landscapes have an environmental value but face threats that need managing.

Fragile tundra ecosystems

Tundra ecosystems (Figure 1) are unique, fragile and sensitive to change. Once damaged, they may never recover.

 See pages 90 and 92 of the student book for more on Arctic and Alpine tundra.

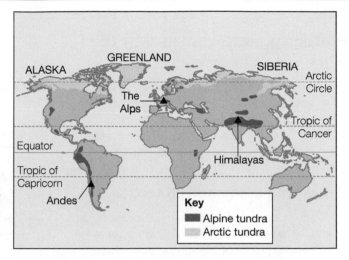

Figure 1 *Tundra systems are found in the Arctic and on mountains close to the snowline – where the climate is cold, rainfall is low and the land is snow-covered for much of the year*

How global warming is changing the Arctic tundra

Rapid warming in the Arctic is causing:

- greater melting of sea ice
- a reduction in snow cover and permafrost
- growth of shrubs and trees that previously could not survive
- **Arctic amplification** – the region is warming twice as fast as the global average.

What is Arctic amplification?

The Arctic gives off more heat than it absorbs. This may be affected by three **positive feedback loops**, which are exacerbating global warming – hence the term '**Arctic amplification**'.

- A reduction in polar sea ice means that a smaller fraction of the sun's radiation is reflected. Darker ice-free water absorbs more solar energy, raising temperatures further.
- When permafrost melts, it releases carbon dioxide (CO_2) and methane (CH_4), increasing the concentration of these gases in the atmosphere (Figure 2).
- Less snow cover means more bare rock is exposed to solar energy. Heat absorption from the sun increases, leading to increased temperatures and snow melt.

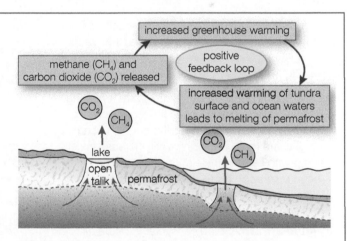

Figure 2 *Melting permafrost, a positive feedback loop*

Changes to the carbon cycle

Not all scientists agree that melting permafrost will release CO_2 and CH_4. Some studies show that, as the permafrost thaws, carbon remains in the soil to be used by new vegetation.

- Warmer temperatures accelerate decomposition (releasing carbon) but also increase the release of nutrients.
- This encourages plant growth and the removal of carbon from the atmosphere through photosynthesis.
- Increased plant growth is a **negative feedback**, opposing Arctic amplification.

Climate change and Alpine ecosystems

Alpine ecosystems are predicted to experience significant change as the climate warms. Vegetation will be pushed higher and some flora and fauna may become extinct. In the Swiss Alps, other possible impacts include:

- glacial lake outburst floods (see Section 2.12)
- the closure of ski resorts below 1500 metres
- increasing melting of permafrost, threatening rock avalanches and mudslides.

Protecting and conserving the Alps

The Alpine Convention is an international treaty. It's an attempt to work on sustainable development within a **legislative (legal) framework**. The Convention covers a range of subjects, including water, soil conservation, landscape protection, transport, energy and climate.

Nature and landscape protection

The Convention includes measures to protect, care for and restore ecosystems as well as preserve the natural environments of animal and plant species.

Water

In the summer, much of Europe's water comes from melting Alpine glaciers. But in less than 100 years, much of the Alps may be ice-free. The **hydrological cycle** (Figure 3) would be affected.

- Snow would decline and rainfall patterns change.
- River discharge patterns would change, with greater flooding in winter and droughts in summer.
- As glaciers melt, water flows will increase initially, along with sediment yield. Further retreat would mean discharge and sediment yield would fall. Water quality would decline, becoming cloudy and sediment-laden.

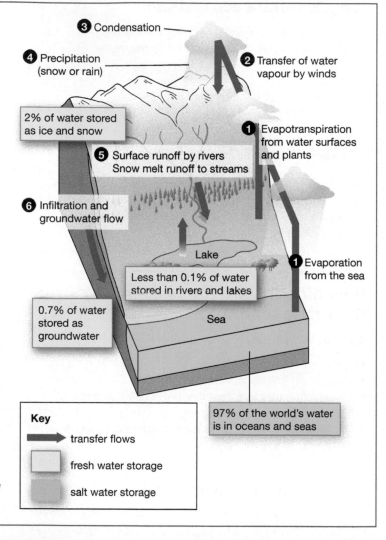

❸ Condensation

❹ Precipitation (snow or rain)

❷ Transfer of water vapour by winds

2% of water stored as ice and snow

❶ Evapotranspiration from water surfaces and plants

❺ Surface runoff by rivers
Snow melt runoff to streams

❻ Infiltration and groundwater flow

Lake

❶ Evaporation from the sea

Less than 0.1% of water stored in rivers and lakes

0.7% of water stored as groundwater

Sea

97% of the world's water is in oceans and seas

Key

 transfer flows

fresh water storage

salt water storage

❯ **Figure 3** *The hydrological cycle. It will be affected by climate change*

An uncertain future

Successful management of tundra regions is challenging and there is a need for co-ordinated global approaches. The 2015 Paris climate agreement (COP21) barely mentioned the Arctic or its indigenous peoples. While developed countries promised to help developing nations with climate **adaptation** and **mitigation**, similar pledges were not made for Arctic communities.

 Ten-second summary

- Tundra is found in glacial and periglacial landscapes.
- Climate change is affecting tundra ecosystems and other natural systems.
- There is a need for global approaches to help manage tundra regions.
- The Alpine Convention is an international treaty designed to protect the natural environment of the Alps while promoting its economic development.

Over to you

List ways in which climate change is affecting *either* the carbon cycle *or* the hydrological cycle in tundra ecosystems. Does this represent a positive or negative feedback system?

Chapter 3
Coastal landscapes and change

What do you have to know?

This chapter studies coastal landscapes and the physical processes that form them. These distinctive landscapes are being changed by physical and human processes, which are affecting their future.

The specification is framed around four enquiry questions:

1 Why are coastal landscapes different, and what processes cause these differences?
2 How do characteristic coastal landforms contribute to coastal landscapes?
3 How do coastal erosion and sea level change alter the physical characteristics of coastlines and increase risks?
4 How can coastal landscapes be managed to meet the needs of all players?

The table below should help you.

- Get to know the key ideas. They are important because 20-mark questions will be based on these.
- Copy the table and complete the key words and phrases by looking at Topic 2B in the specification. Section 2B.1 has been done for you.

Key idea	Key words and phrases you need to know
2B.1 The coast, and wider littoral zone, has distinctive features and landscapes.	littoral zone; backshore, nearshore and offshore zones; coastal types, dynamic zone, changes of sea level; inputs from rivers, waves and tides; rocky coasts (high and low relief), high-energy environment, coastal plain landscapes (sandy and estuarine), supplies of sediment from terrestrial and offshore sources; low-energy environment
2B.2 Geological structure influences the development of coastal landscapes at a variety of scales.	
2B.3 Rates of coastal recession and stability depend on lithology and other factors.	
2B.4 Marine erosion creates distinctive coastal landforms and contributes to coastal landscapes.	
2B.5 Sediment transport and deposition create distinctive landforms and contribute to coastal landscapes.	
2B.6 Subaerial processes of mass movement and weathering influence coastal landforms and contribute to coastal landscapes.	
2B.7 Sea level change influences coasts on different timescales.	
2B.8 Rapid coastal retreat causes threats to people at the coast.	
2B.9 Coastal flooding is a significant and increasing risk for some coastlines.	
2B.10 Increasing risks of coastal recession and coastal flooding have serious consequences for affected communities.	
2B.11 There are different approaches to managing the risks associated with coastal recession and flooding.	
2B.12 Coastlines are now increasingly managed by holistic integrated coastal zone management (ICZM).	

You need to know:

- how geology affects the coast
- about the littoral zone
- how coasts can be classified.

Big idea

The coast and littoral zone consist of distinctive features and landscapes.

The role of geology

The UK's coastline varies, from the rocky coasts of Cornwall to the low-lying muddy estuarine coast of The Wash. **Geology** plays a major role in this.

See Figure 1 on page 97 of the student book for a map of UK geology and photos showing how coastal landscapes vary.

Resistant rock coastlines

Due to its geology, Cornwall's **rocky coastline** can withstand winter storms without suffering rapid erosion. Like the rest of western and northern Britain, much of Cornwall consists of older rocks that are resistant to the erosive power of the sea, including:

- **igneous** rocks (e.g. granite)
- older compacted **sedimentary** rocks (e.g. old red sandstone)
- **metamorphic** rocks (e.g. slates and schists).

Coastal plain landscapes

Britain's eastern and southern coasts consist of areas of younger, much weaker sedimentary rocks (including chalks, clays, sand and sandstone).

- The Wash is an area of **low, flat relief** – referred to as a **coastal plain**.
- Much of the coast of eastern England consists of low-lying **sandy** beaches, e.g. Skegness on the Lincolnshire coast.

High-energy coastlines

Rocky coasts are generally found in **high-energy environments**. In the UK, these tend to be:

- stretches of the Atlantic-facing coast (with powerful waves)
- where the rate of erosion exceeds the rate of deposition.

Erosional landforms are found in these environments.

Low-energy coastlines

Sandy and estuarine coasts are generally found in **low-energy environments**. In the UK, these tend to be:

- stretches of the coast where waves are less powerful or where it is sheltered
- where the rate of deposition exceeds the rate of erosion.

Depositional landforms are found in these environments.

The coast as a system

The coast is constantly changing, so it helps to think of it as a system driven by wave energy. The inputs, processes and outputs are linked (Figure 1). Any change to one component impacts on the rest of the system.

Inputs
- Marine – waves, tides, storm surges
- Atmospheric – weather/climate, climate change, solar energy
- Land – rock type and structure, tectonic activity
- People – human activity, coastal management

Processes
- Weathering
- Mass movement
- Erosion
- Transport
- Deposition

Outputs
- Erosional landforms
- Depositional landforms
- Different types of coasts

▲ *Figure 1 The coastal system*

The littoral zone

The **littoral** (coastal) **zone** stretches into the sea and onto the shore. It is constantly changing because of the dynamic interaction between the processes operating in the seas and oceans and on land. It varies because of:

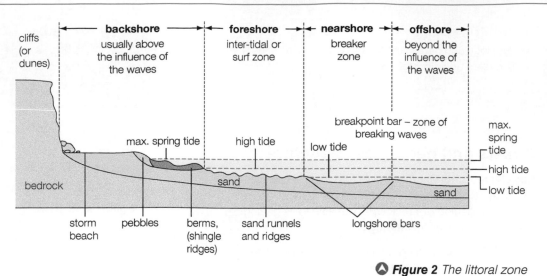

- short-term factors (e.g. waves, tides, storms)
- long-term factors (e.g. changes to sea levels, climate change).

The littoral zone consists of sections (Figure 2). The backshore and foreshore are where most human activity occurs and where most physical processes (erosion, deposition, transport and mass movement) operate.

Figure 2 *The littoral zone*

Sediment supply

The processes of weathering and erosion produce sediment, which is transported and deposited to produce coastal landforms. In The Wash (Figure 3), sediment originates from:

- cliffs eroding between West Runton and Weybourne
- **tidal currents**, which pick up glacial deposits from the shallow sea floor
- erosion of the Holderness cliffs – sediment is carried southwards in suspension
- sand carried southwards along the Lincolnshire coast
- rivers discharging into The Wash.

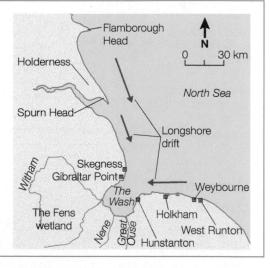

 Figure 3 *The sources of sediment for The Wash*

Classifying coasts

Coasts can be classified depending on:

- **geology** – creating rocky, sandy and estuarine coasts, plus concordant and discordant coasts (see Section 3.2)
- the level of **energy** – creating high- or low-energy coasts
- the **balance** between erosion and deposition – creating either erosional or depositional coasts (see Sections 3.4 and 3.5).
- changes in **sea level** – creating emergent or submergent coasts (see Section 3.7).

Ten-second summary

- Geology plays a major role in producing distinctive coastal landscapes.
- Rocky coasts are generally found in high-energy environments, and sandy and estuarine coasts in low-energy environments.
- The coast is a system with inputs, processes and outputs.
- The littoral zone is a dynamic zone that changes rapidly.
- Coasts can be classified in different ways.

Over to you

Distinguish between the following pairs of terms:

a geology and relief
b high- and low-energy environments
c littoral zone and coastline.

Student Book
See pages 100–105

You need to know:

- how geology and geological structure influences coastal morphology, erosion rates and cliff profiles.

Big idea

Geology and geological structure influence the development of coastal landscapes and rates of recession.

Coasts and geological structure

Coastal morphology (the shape and form of coastal landscapes and their features) is related not only to the underlying geology, or rock type, but also to its geological structure (known as its **lithology**).

The **relief** is also affected by geology. There is a direct relationship between rock type, geological structure (lithology) and **cliff profiles** (Figure 1).

Lithology means any of the following characteristics:

- **strata**
- **bedding planes**
- **joints** or cracks
- **folds**
- **faults**
- **dip**.

a) horizontal strata produce steep cliffs

----- bedding planes

b) rocks dip gently towards the sea, with almost vertical joints

joints opened by weathering and pressure release

c) steep dip towards the sea

rock slabs slide down the cliff along bedding planes

d) rocks dip inland producing a stable, steep cliff profile

e) rocks dip inland but with well-developed joints at right angles to bedding planes

joints act as slide planes

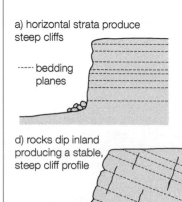

◀ **Figure 1** *Cliff profiles and geological structure*

Concordant coasts

Along the south coast of the Isle of Purbeck, the bands of more- and less-resistant rock run parallel to the coast, forming a **concordant coast**.

Figure 2 shows how the geology and geological structure of the Isle of Purbeck has influenced the coastal morphology.

See page 103 of the student book for information on the Dalmatian and Haff coasts, both types of concordant coastlines.

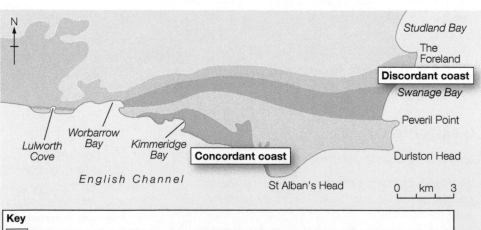

Key

- Bagshot and Tertiary beds – unconsolidated sands and clays, less resistant to erosion and forms bays, e.g. Studland.
- Chalk – strong and resistant to erosion. Has formed cliffs and a headland - The Foreland.
- Wealden beds – mainly unconsolidated clay. Less resistant to erosion. Has led to the formation of Swanage Bay.
- Purbeck and Portland beds – mainly limestone. Resistant and has led to the creation of the headlands, e.g. Peveril Point. Jointed, which has created lines of weakness.
- Kimmeridge clay – relatively unconsolidated.

◢ **Figure 2** *The geology of the Isle of Purbeck*

Discordant coasts

Along the eastern coast of the Isle of Purbeck, the bands of more- and less-resistant rock run at right angles to the coast, south from Studland Bay to Durlston Head, forming a **discordant coast**. The more-resistant rocks (folded into ridges) emerge at the coast as headlands and cliffs, whilst less-resistant rocks form bays (Figure 2).

Geology and rates of coastal recession

The geology and lithology at the coast affect the speed at which it erodes or **recedes**.

- Igneous rocks (e.g. granite) are crystalline, resistant and **impermeable**.
- Sedimentary rocks are formed in strata (layers). Jointed sedimentary rocks (e.g. sandstone and limestone) are **permeable**. Other sedimentary rocks (e.g. chalk) have air spaces between particles, making them **porous**. Shale is fine-grained and compacted, making it impermeable.
- Metamorphic rocks (e.g. marble) are hard, impermeable and resistant.
- Unconsolidated materials (e.g. boulder clays of the Holderness coast) are loosely compacted and therefore easily eroded.

Processes, such as weathering and mass movement, also affect the rate of erosion or recession.

Headlands and bays

Headlands jut out into the sea, with bays lying between them. Headlands and bays commonly form when rocks of different strengths are exposed at the coast (Figure 3). More-resistant rocks, such as chalk and limestone (or igneous and metamorphic rocks), tend to form headlands, whilst weaker rocks (e.g. shale and clays) are eroded to form bays.

- As waves approach a headland, the depth of water decreases. Velocity reduces due to friction, so that waves compress, becoming higher, steeper and closer together. Their erosive power increases.
- When waves enter a bay, the water is deeper. The waves lose velocity less rapidly, and are lower and less steep than those off the headland. This allows deposition to take place.

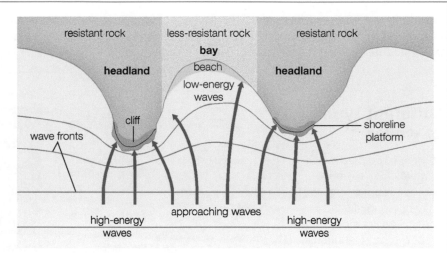

Figure 3 Headlands and bays

Ten-second summary

- Geological structure influences coastal morphology and rates of coastal recession.
- Concordant coasts are formed where bands of more- and less-resistant rock run parallel to the coast.
- On discordant coastlines, bands of more- and less-resistant rock run at right angles to the coast.
- Headlands and bays commonly form on discordant coastlines.

Over to you

Create a spider diagram to show how geology and geological structure influence coastal morphology, erosion rates and cliff profiles.

You need to know:

- about different types of waves
- how they influence beach morphology and profiles.

Big idea

Different types of waves influence beach morphology and profiles.

What causes waves?

Most originate locally, forming when wind blows over water.

- Their size is related to wind speed.
- They build up over time.
- Wind creates frictional drag, which produces movement in the upper surface of the water.
- Water particles move in a circular orbit as waves move (or ripple) across the surface.

Swell waves

Some waves originate in mid-ocean. The distance of water over which they move is called the **fetch**. On the UK coast, mid-ocean **swell waves** appear as larger waves amongst smaller locally generated waves.

See page 106 of the student book for what happens when waves approach the coast.

Different types of waves

There are two main types of waves: **constructive** and **destructive** (Figure 1). They have significant impacts on coastal processes and landforms.

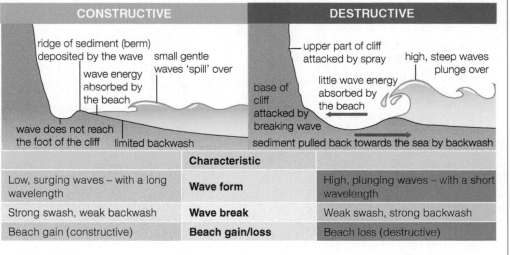

▶ **Figure 1** *Constructive and destructive waves*

	Characteristic	
Low, surging waves – with a long wavelength	**Wave form**	High, plunging waves – with a short wavelength
Strong swash, weak backwash	**Wave break**	Weak swash, strong backwash
Beach gain (constructive)	**Beach gain/loss**	Beach loss (destructive)

Beach morphology and profiles

Beaches consist of loose material, so their **morphology** (form or shape) alters as waves change. The material along a beach profile also varies in **size** and **type**, depending on distance from the shoreline.

Beach **profiles** are steeper in summer.

- The swash of constructive waves deposits larger material at the top of the beach, creating a **berm**.
- As the berm builds, the backwash becomes weaker, so beach material becomes smaller towards the shoreline.

In winter, berms are eroded by plunging destructive waves.

- Strong backwash transports sediment offshore (depositing it as **offshore bars**).
- The backwash can exert an undertow, dragging sediment back as the next wave arrives over the top.

Figure 3 on page 107 of the student book shows how pebble size and shape change along a beach transect.

 Ten-second summary

- Most waves originate locally.
- Waves that originate in mid-ocean are larger swell waves.
- There are two main types of waves: constructive and destructive.
- Beach profiles are steeper in summer and gentler in winter.

Over to you

Design a flow diagram to show what happens as waves approach the coast and eventually break.

You need to know:
- about coastal erosion processes
- how the processes of coastal erosion create distinctive landforms.

Big idea

Marine erosion creates distinctive coastal landforms, contributing to coastal landscapes.

Cliff-foot erosion

Waves erode the cliff foot in the following ways.

- **Abrasion** – also known as **corrasion** (Figure 1). The size and amount of sediment picked up by waves, along with the wave type, determines the relative importance of abrasion.
- **Hydraulic action** (Figure 2).
- **Corrosion** – When cliffs are formed from alkaline rocks (e.g. chalk) or alkaline cement bonds rock particles together, weak acids in seawater can dissolve them.
- **Attrition** – that is, the gradual reduction of rock particles by impact and abrasion, as they are moved by waves, tides and currents. It gradually reduces particle size and makes sediment rounder/smoother.

Each erosive process can cause cliffs to become undercut and unstable, leading to their collapse and retreat.

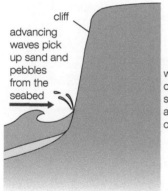

cliff
advancing waves pick up sand and pebbles from the seabed

wave breaks at base of cliff, throwing sediment at it, known as abrasion or corrasion

🔺 **Figure 1** *Cliff-foot erosion: abrasion*

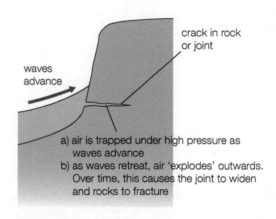

waves advance

crack in rock or joint

a) air is trapped under high pressure as waves advance
b) as waves retreat, air 'explodes' outwards. Over time, this causes the joint to widen and rocks to fracture

🔺 **Figure 2** *Cliff-foot erosion: hydraulic action*

How waves and lithology influence erosion

The rate and type of erosion experienced on the coast is influenced by the size and type of waves.

- Most erosion happens during winter when destructive waves are large and powerful.
- Hydraulic action and abrasion attack differences in rock resistance (or weaknesses, e.g. cracks, joints).

Lithology influences erosion.

- At a small scale, geological weaknesses are eroded more quickly.

- Bands of more-resistant rock between weaker joints and cracks erode more slowly. The selective erosion of areas of weakness is called **differential erosion**.
- At a medium and larger scale, areas of resistant rock generally form cliffs and headlands, while weaker rocks form lowland areas with bays and inlets.

Generally, erosion is faster where the rocks forming the coastline are weaker.

- At Holderness, weak boulder clays have eroded inland by 120 metres in a century.
- Resistant granites at Land's End have only eroded by 10 cm in the same period.

Landforms created by coastal erosion

Wave-cut notches and shoreline platforms

When waves break against the foot of a cliff, erosion is concentrated close to the high-tide line, creating a **wave-cut notch**, which begins to undercut the cliff.

- As the notch gets bigger, the rock above becomes unstable and eventually the upper part of the cliff collapses.
- As the processes are repeated, the notch migrates inland and the cliff retreats (Figure 3), leaving a remnant behind as a **shoreline platform** (wave-cut platform).

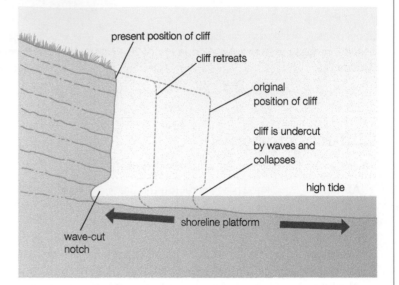

 Figure 3 Cliff retreat

Cliffs

The steepest cliffs are found where rock strata are vertical or horizontal, or have almost vertical joints. The gentlest are found where rock dips towards or away from the sea (see Figure 1 in Section 3.2).

Constant wave action and erosion against the base of the cliff ensures that it maintains its profile as it retreats inland (Figure 3).

Caves, arches, stacks and stumps

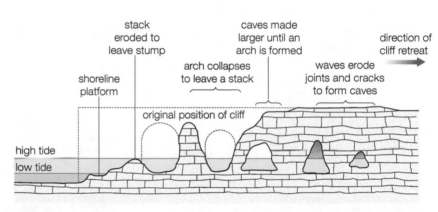

 Figure 4 The formation of caves, arches, stacks and stumps at a headland

- The erosion of rocks like limestone exploits lines of weakness to form distinctive features (Figure 4).
- When joints and faults are eroded by hydraulic action and abrasion, this can create **caves**.
- If two caves on both sides of a headland join, or a single cave is eroded through a headland, an **arch** is formed. The gap is further enlarged by erosion and weathering, becoming wider at the base.
- Eventually, the top of the arch will become unstable and collapse, leaving an isolated pillar of rock called a **stack**. The stack will continue to be eroded by the sea. As it collapses and is eroded further, it may only appear above the surface at low tide and is known as a **stump**.

Ten-second summary

- Waves attack the cliff base and erode the cliff foot by the processes of abrasion, hydraulic action, corrosion and attrition.
- Wave type, size and lithology influence coastal erosion.
- Coastal erosion creates a range of distinctive landforms.

Over to you

Draw a flow chart to show the development of wave-cut notches, shoreline platforms and cliffs.

You need to know:
- about sediment transport
- how transport and deposition processes produce coastal landforms that can be stabilised by plant succession
- how sediment cells work.

Big idea
Sediment transport and deposition create distinctive landforms and contribute to coastal landscapes.

Longshore drift

Figure 1 shows the process of **longshore drift** where sediment is transported along the coast.

▶ **Figure 1**
The process of longshore drift (1–4). When the removal of sediment is greater than the supply of new sediment, the beach is eroded

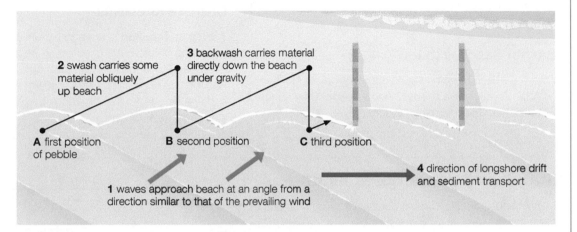

2 swash carries some material obliquely up beach

3 backwash carries material directly down the beach under gravity

A first position of pebble

B second position

C third position

4 direction of longshore drift and sediment transport

1 waves approach beach at an angle from a direction similar to that of the prevailing wind

Marine transport

There are four main methods of coastal sediment transport:

- **traction**
- **saltation**
- **suspension**.
- **solution**.

Tides and currents

Tides and currents affect longshore drift.

- **Tides** are caused by the gravitational pull of the moon and, to a lesser extent, the sun. The UK coastline has two high and two low tides a day.
- The relative difference in height between high and low tides is the **tidal range**. A high tidal range creates relatively powerful tidal currents, important in transporting sediment.

Coastal depositional landforms

Deposition occurs when waves no longer have enough energy to transport sediment.

Spits

A spit is a long narrow feature made of sand or shingle. Figure 2 shows how they form.

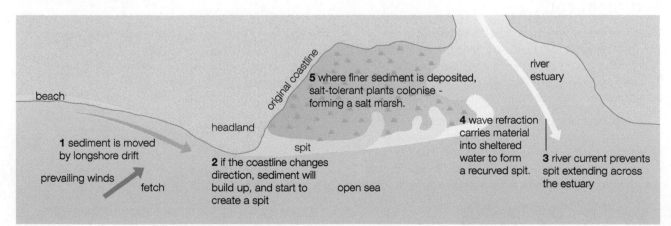

original coastline

river estuary

5 where finer sediment is deposited, salt-tolerant plants colonise - forming a salt marsh.

beach

headland

1 sediment is moved by longshore drift

prevailing winds

fetch

spit

2 if the coastline changes direction, sediment will build up, and start to create a spit

open sea

4 wave refraction carries material into sheltered water to form a recurved spit.

3 river current prevents spit extending across the estuary

▲ **Figure 2** *The formation of a spit (1–5)*

Beaches

Beaches are commonly found in bays. Wave refraction creates a low-energy environment leading to deposition. Beaches can be sand or shingle, depending on the nature of the sediment and the power of the waves. Beaches can be **swash-aligned** or **drift-aligned**.

Offshore bars

Offshore bars (or **sandbars**) are submerged / partly exposed ridges of sand / coarse sediment. Destructive waves erode sand from the beach and deposit it offshore in bars.

Barrier beaches (bars)

Where a beach or spit extends across a bay to join two headlands, it forms a **barrier beach** or **bar**. They can trap water behind them forming lagoons.

For more on wave refraction leading to deposition, see page 105 of the student book. See page 114 of the student book for data on sediment characteristics along a drift-aligned beach and for Figure 5, showing how swash-aligned and drift-aligned beaches form.

Tombolos

A **tombolo** is a beach (or ridge of sand and shingle) that forms between a small island and the mainland. They may be covered at high tide.

Cuspate forelands

A **cuspate foreland** is a triangular-shaped headland extending out from the coastline. It occurs where a coast is exposed to longshore drift from opposite directions. Sediment is deposited where the two meet, forming a triangular shape. As vegetation begins to grow, it stabilises the sediment.

Stabilising depositional landforms

Many depositional landforms consist of sediments that can be easily eroded and transported.

- **Dunes** can develop where sand is trapped by debris towards the back of the beach.
- **Salt marshes** are areas of silty sediments that accumulate around estuaries or lagoons.
- Vegetation helps to stabilise dunes and salt marshes as a result of **plant succession** (Figure 3).

For more on plant succession and sand dunes, see page 116 of the student book.

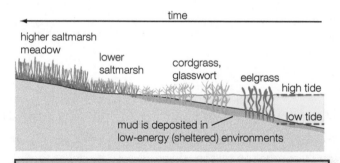

- As mud flats develop, **halophytes** (salt-tolerant plants, e.g. eelgrass) begin to colonise them.
- Other halophytes (e.g. cordgrass) help to slow tidal flow and trap more sediment.
- As sediment accumulates, the surface becomes drier and plants begin to colonise (e.g. sea asters).

 Figure 3 *Salt marsh formation and plant succession*

Sediment cells

Sediment moves along the coast in **sediment cells** between beaches, cliffs and sea through the processes of erosion, transport and deposition. The coastline of England and Wales is divided into 11 major sediment cells, though smaller ones can operate within them.

Sediment cells act as **systems** with **sources, transfers** (or **flows**) and **sinks**. The amount of sediment available within a cell is the **sediment budget**. Within each cell, depositional features build up which are in **equilibrium** with the amount of sediment available.

For information on feedback in sediment cells, see page 117 of the student book.

Ten-second summary

- Sediment transport is influenced by wave angle, tides, currents and longshore drift.
- Transport and deposition produce distinctive coastal landforms.
- Vegetation can stabilise depositional landforms as a result of plant succession.
- Sediment cells act as systems.

 Over to you

Create a set of stickie notes on how each coastal depositional landform is formed. Add an example (check the student book for these).

You need to know:

- about the importance of weathering in producing sediment
- that mass movement creates distinctive landforms at the coast.

Big idea

Sub-aerial processes influence coastal landforms and contribute to coastal landscapes.

Sub-aerial processes

Weathering and mass movement are **sub-aerial processes**.

- **Weathering** is the gradual breakdown of rock, *in situ,* at or close to the ground surface. There are three types (see coloured panels). Weathering creates sediment, which the sea uses to help erode the coast.
- **Mass movement** is the movement of weathered material downslope under gravity. It can reshape the coastline, e.g. through slumped cliffs.

Mechanical weathering

- **Freeze-thaw weathering** (frost-shattering) occurs when water enters cracks/joints in rock (Figure 1). As water freezes, it expands by about 10%, exerting pressure, forcing cracks to widen. Repeated freezing and thawing causes fragments of rock to break away and collect at the base of the cliff as **scree**.
- **Salt-weathering** – When salt water evaporates, it leaves salt crystals behind. These can grow and exert stresses in the rock, causing it to break up. Salt can also corrode rock.
- **Wetting and drying** – Rocks rich in clay (e.g. shale) expand when they get wet and contract as they dry. This can cause them to break up.

Biological weathering

This occurs in several ways:

- Plant roots grow into cracks in a cliff face. Cracks widen as the roots grow, breaking up the rock (Figure 2).
- Water running through decaying vegetation becomes acidic, leading to increased chemical weathering.
- Birds and animals dig burrows into cliffs.
- Marine organisms can burrow into rocks or secrete acids.

Chemical weathering

Carbonation – Rainwater absorbs carbon dioxide from the air, forming weak carbonic acid. This reacts with calcium carbonate in rocks to form calcium bicarbonate, which is easily dissolved.

⚠ **Figure 1** *A major rock fall at the White Cliffs of Dover in February 2001, caused by freeze-thaw weathering*

⚠ **Figure 2** *Biological weathering*

Mass-movement processes

Mass movement can be classified in different ways (Figure 3). The movement depends on different factors, e.g. angle of the slope/cliff; rock type and structure; vegetation cover; how wet the ground is.

⚠ **Figure 3** *Classifying different types of mass movement. In a slide, the material remains intact (moving 'en masse'). In a flow, the material becomes jumbled up*

Nature of movement	Rate of movement	Type of mass movement
Flow	Imperceptible	• Soil creep • Solifluction
Flow	Slow to rapid	• Earth flow/mudflow
Slide	Slow to rapid	• Rock/debris fall • Rock/debris slide • Slump

Mass movement – flows

Soil creep

- Very slow downhill movement of soil particles (Figure 4).
- This is the slowest form of mass movement.

Solifluction

- Occurs mainly in tundra areas. When the top layer of soil thaws and becomes saturated, it flows over the frozen layer beneath.
- Averages 5 cm to 1 metre a year.

Earth flows and mudflows

- An increase in the amount of water (e.g. due to heavy rain) can reduce friction, causing earth and mud to flow over bedrock (Figure 5).

Mass movement – slides

Rock falls

- Most likely to occur when strong, jointed, steep rock faces/cliffs are exposed to mechanical weathering (e.g. freeze-thaw).
- Material falls to form **scree** (**talus**) at the foot of the slope/cliff (Figure 6).
- Block falls are similar. A large block of rock falls from the cliff as a single piece due to the jointing of the rock.

Rock/debris slides

- Rocks that are jointed, or have bedding planes roughly parallel to the slope or cliff surface, are susceptible to landslides (Figure 7).
- An increase in the amount of water can reduce friction, causing sliding.
- Slabs of rock/blocks can slide along a slide or slip plane.

Slumps

- Often occur in saturated conditions on moderate to steep slopes (Figure 8).
- There is a rotational movement.
- Common where softer materials (clays or sands) overlie more-resistant or impermeable rock (e.g. limestone or granite).
- Slumping causes **rotational scars** and **terraced cliff profiles**.

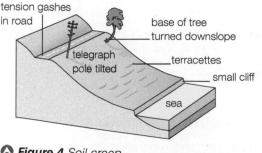

▲ *Figure 4* Soil creep

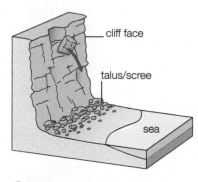

▲ *Figure 5* A mudflow

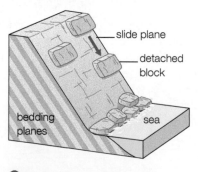

▲ *Figure 6* A rock fall

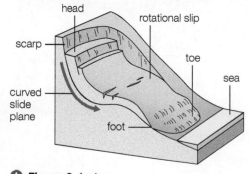

▲ *Figure 8* A slump

▲ *Figure 7* A rock/landslide

Ten-second summary

- Weathering creates sediment, which the sea uses to help erode the coast.
- Weathering can be mechanical, biological or chemical.
- Mass movement can be classified as flows or slides, which occur at different speeds.
- Mass movement creates distinctive landforms.

Over to you

Distinguish between these pairs of terms:

a weathering and mass movement
b flows and slides
c slides and slumps
d soil creep and solifluction
e rockfalls and slumping.

You need to know:

- that contemporary sea level change can result from global warming and tectonic activity
- that longer-term sea level change results from eustatic and isostatic changes, and tectonic activity
- that sea level change has produced emergent and submergent coastlines.

Big idea

Sea level change influences coasts over different timescales.

Contemporary sea level change

Global warming

Kiribati consists of 33 islands in the Pacific Ocean. They lie one metre or less above sea level, which, in places, is rising by 1.2 cm a year (four times faster than the global average) due to global warming.

- Melting polar ice sheets (and glaciers) and **thermal expansion** are causing levels to rise.
- Scientists estimate that, by 2100, average sea levels will have risen by 30–100 cm.
- Low-lying nations, like Kiribati, are at risk.

If the islands are submerged, Kiribati's population could become **environmental refugees** (people forced to migrate due to environmental change).

Tectonic activity

The 2004 Boxing Day earthquake caused a tsunami that temporarily flooded the city of Banda Aceh on Sumatra, Indonesia. It also caused the Earth's crust at Banda Aceh to sink, *permanently* flooding parts of the city.

For more on the 2004 earthquake and its impact on sea level, see page 124 of the student book.

Longer-term sea level change

Eustatic change and isostatic change

Sea level is measured relative to land. The two types of sea level change are:

- **Eustatic change** – when the sea level itself rises or falls.
- **Isostatic change** – when land rises or falls, relative to the sea.

Eustatic change is global.

- In glacial periods, precipitation falls as snow, forming ice sheets that store water normally held in the oceans. Sea levels fall.
- At the end of glacial periods, ice sheets begin to melt.
- Stored water flows into rivers and the sea, and sea levels rise.

Isostatic change occurs locally (Figure 1).

- During glacial periods, the weight of ice sheets makes the land sink (**isostatic subsidence**).
- As ice begins to melt at the end of a glacial period, its reduced weight causes the land to readjust and rise (called **isostatic recovery**).

Past tectonic activity

Past tectonic activity has had a direct impact on some coasts and sea levels, due to:

- the uplift of mountain ranges and coastal land at destructive and collision plate margins
- local tilting of land.

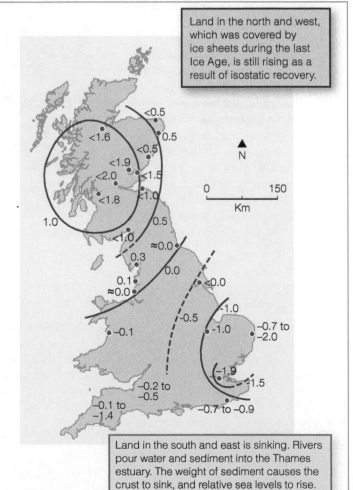

Land in the north and west, which was covered by ice sheets during the last Ice Age, is still rising as a result of isostatic recovery.

Land in the south and east is sinking. Rivers pour water and sediment into the Thames estuary. The weight of sediment causes the crust to sink, and relative sea levels to rise.

Figure 1 *Two types of isostatic change that occurred in the UK since the last Ice Age, showing how much parts of the UK are either rising or falling in millimetres*

Landforms caused by changing sea level

Emergent coastline landforms

A fall in sea level exposes land previously covered by the sea, creating an **emergent coastline**.

- **Raised beaches** are common on the west coast of Scotland. As land rose due to isostatic recovery, former shoreline platforms and their beaches were raised above the present sea level.
- The remains of eroded cliff lines (**relic cliffs**) can be found behind raised beaches, with wave-cut notches and caves – evidence of past marine erosion.

Submergent coastline landforms

A rise in sea level floods the coast, creating a **submergent coastline**.

- **Rias** are winding inlets with irregular shorelines.
  They form when valleys in a dissected upland area are flooded. They are common in south-west England (Figure 2), where sea levels rose after the last Ice Age.
- **Dalmatian coasts** are similar to rias. Rivers flow almost parallel to the coast rather than at right angles to it.
- **Fjords** are formed when deep glacial troughs (see Section 2.8) are flooded by rising sea levels. They are long and steep-sided, with a U-shaped cross-section (Figure 3).

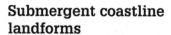 **Figure 2** *The Kingsbridge estuary, a ria in Devon*

Figure 3 *Milford Sound, a fjord in New Zealand's South Island*

 Ten-second summary

- Contemporary sea level change poses a risk to some low-lying coastal areas.
- Longer-term sea level change can be eustatic or isostatic.
- Falls in sea level create emergent coastlines with associated landforms.
- Rises in sea level flood the coast, creating submergent coastlines with associated landforms.
- Tectonic activity can affect coasts and sea levels.

 Over to you

Explain the difference between eustatic and isostatic change.

See pages 126–131

You need to know:

- what causes rapid coastal retreat
- that rates of retreat are influenced by a range of factors including subaerial processes
- about the economic and social losses of coastal retreat.

Big idea

Rapid coastal retreat causes threats to people at the coast.

Erosion on the Holderness coast

Holderness stretches between Bridlington and Flamborough Head in the north and Spurn Head in the south (Figure 1). On average, it's eroding nearly 2 metres a year. This rapid retreat is caused by geology, fetch, and longshore drift and beach material.

Geology

- Most of the coast consists of **boulder clay.** It has little resistance to erosion and produces shallow, sloping cliffs.
- The boulder clay is surrounded by chalk, which has created a headland at Flamborough Head. Erosion has created cliffs, arches and stacks there.

Fetch

Holderness is exposed to winds and waves from the north-east, with a small fetch (about 500–800 km across the North Sea). The waves attacking the coast are influenced by other factors, which increase their size and power.

- Currents (or **swell**) circulate around the UK from the Atlantic Ocean into the North Sea, adding energy to the waves. Powerful **destructive waves** often attack the coastline.
- Low-pressure **weather systems** and winter storms passing over the North Sea are often intense, producing strong winds, waves and high tides.
- The **sea floor** is relatively deep along the Holderness coast. Waves reach the cliffs without being weakened by friction with shallow beaches.

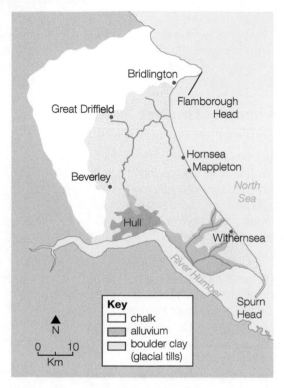

Key
- chalk
- alluvium
- boulder clay (glacial tills)

▲ *Figure 1 The geology of the Holderness coast*

Longshore drift and beach material

- Boulder clay erodes to produce clay particles, which are transported out to sea. Narrow beaches offer little friction to absorb wave energy.
- Tides flow southwards, transporting sand by **longshore drift**, leaving the cliffs at Holderness poorly protected against wave attack.

Sub-aerial processes and coastal erosion

The cliffs at Holderness are affected by sub-aerial processes.

- The main types of weathering experienced are freeze-thaw and alternate wetting and drying of the boulder clay, making it crumbly in dry periods.
- **Slumping** affects the boulder clay cliffs. Alternate wetting and drying causes expansion and shrinkage, producing cracks. Rain entering cracks percolates into the cliff, which becomes lubricated and heavier. The weakened cliff cannot support the extra weight, and the clay slides downslope. Slumped material collects at the cliff base and is removed by the sea, so the cliff line retreats.

Human actions and coastal retreat

The actions of people and organisations (**players**) can impact on coastal retreat.

Key players on the Holderness coast

- **Central and local government agencies** – The Environment Agency and local authorities are jointly responsible for coastal management. Funding for both has been cut.

- **Stakeholders in the local economy** – The tourist industry, farmers and residents want greater coastal protection. Insurance companies are increasingly refusing to insure vulnerable properties.

- **Environmental stakeholders** – English Nature and the RSPB want to protect Spurn Head. A continuing flow of sand southwards by longshore drift is essential.

The impact of coastal management

The gaps on the graph (Figure 2) show where coastal defences are preventing erosion on the Holderness coast. However, higher rates of erosion occur immediately south of those defences.

- A sea wall, rock armour and groynes help to protect Hornsea but interrupt the flow of beach material by longshore drift.
- The beach at Mappleton is starved of material and its cliffs are exposed to greater wave attack. This is called **terminal groyne syndrome**.

 Figure 11 on page 130 of the student book shows coastal retreat at Hornsea since 1854.

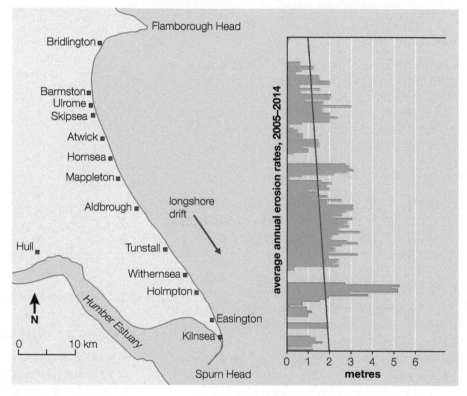

 Figure 2 Erosion rates opposite the locations along the Holderness coast that they relate to, 2005–2014

Holderness – economic and social losses

Economic and social losses	East Riding Coastal Change Fund
<ul><li>Tourist industry – Chalets fall into the sea (so fewer people visit).</li><li>200 homes will be lost in the next 80 years as the coast retreats.</li><li>Infrastructure is lost to the sea.</li><li>There is no compensation for loss of private property and land due to coastal erosion in England.</li></ul>	<ul><li>Offers limited help to those affected by coastal erosion:<ul><li>**Relocation packages** fund property demolition and some relocation costs.</li><li>**Adaptation packages** can help towards the cost of replacing property/business.</li><li>Assistance grants help to adapt properties at risk from future erosion.</li></ul></li></ul>

 Figure 3 Losses due to coastal erosion, and help available in Holderness

Ten-second summary

- Physical and human factors influence coastal retreat.
- Sub-aerial processes affect rates of coastal retreat.
- Coastal retreat has serious consequences for coastal communities.

Over to you

Draw a spider diagram to show the physical and human factors affecting coastal retreat at Holderness.

You need to know:

- that local factors can increase flood risk on low-lying and estuarine coasts
- how storm surge events can cause severe local flooding
- that climate change can increase coastal flood risks.

Big idea

Coastal flooding is a significant and increasing risk for some coastlines.

Developing country – coastal flooding and storm surges

Cyclone Sidr swept into Bangladesh from the Bay of Bengal in November 2007, bringing a **storm surge** up to 6 metres high, heavy rain and strong winds of up to 223 km/hr. Coastal districts and offshore islands suffered the most deaths and worst effects (Figure 1).

Why is Bangladesh at risk of flooding?

Bangladesh is threatened frequently by river and coastal flooding.

- It is the world's most densely populated country.
- 46% of the population lives in places less than 10 metres above sea level.
- It lies on the floodplains of three major rivers, which, together with smaller rivers, empty into the Bay of Bengal through a series of estuaries.

Increasing flood risk

- **Subsidence** – Clearing and draining land for cultivation, plus building large earth embankments, has caused some estuarine islands to shrink and subside by up to 1.5 metres in the last 50 years.
- **Removing vegetation** – 70% of Bangladesh's mangrove-forested coastline is retreating by up to 200 metres annually due to erosion, rising sea levels and removal of mangrove vegetation. Mangrove forests provide protection and shelter against storm winds, floods and tsunami by absorbing and dispersing tidal surges.

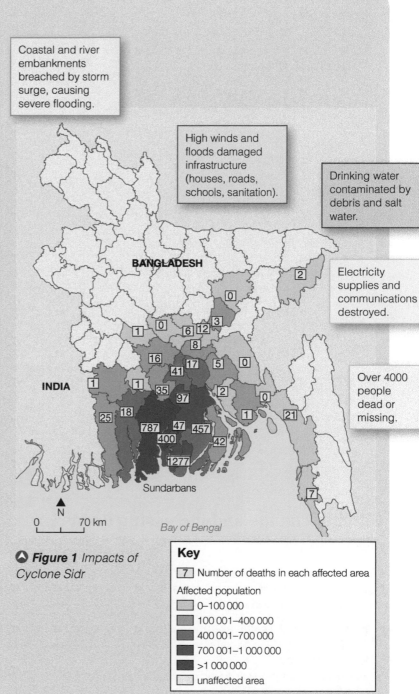

Coastal and river embankments breached by storm surge, causing severe flooding.

High winds and floods damaged infrastructure (houses, roads, schools, sanitation).

Drinking water contaminated by debris and salt water.

Electricity supplies and communications destroyed.

Over 4000 people dead or missing.

▲ **Figure 1** *Impacts of Cyclone Sidr*

Key

| 7 | Number of deaths in each affected area |

Affected population
- 0–100 000
- 100 001–400 000
- 400 001–700 000
- 700 001–1 000 000
- >1 000 000
- unaffected area

Storm surges

Storm surges are changes in sea level caused by intense low-pressure systems (**depressions** and **tropical cyclones**) and high wind speeds. For every decrease in air pressure of 10 mb, sea level rises by 10 cm. During tropical cyclones, air pressure may be 100 mb lower than normal, raising sea level by 1 metre.

Developed countries – coastal flooding and storm surges

Over the winter of 2013/14, the UK experienced a succession of major storms. One of the worst occurred during 5–6 December, bringing a storm surge that affected the east coast of the UK and countries in northern Europe (Figure 2). The surge was caused by:

- Intense low pressure.
- Sea shape and coastline – The North Sea is open to the Atlantic Ocean and tapers southwards in a funnel shape, allowing strong northerly winds to push storm surges south.
- Sea depth – The North Sea gets shallower and narrower towards the south, increasing the height of tides and storm surges.
- High seasonal tides.

Figure 2 *Storm and storm surge across northern Europe, 5–6 December 2013*

Impacts of the storm

- On the East Frisian coast in Germany and on the Dutch border, the storm surge reached 3.7 metres above sea level.
- Gusts of over 200 km/hr in Scotland.
- The Thames Barrier closed to protect London and the Eastern Scheldt storm-surge barrier in the Netherlands closed.
- At Hemsby in Norfolk, cliff erosion resulted in properties collapsing into the sea.
- 1400 homes flooded in the UK.
- Rail services in eastern England were disrupted.
- Forced evacuation along the coast of eastern England and northern Wales.

Key

Hazards
- ◉ storm surge
- ★ strong wind
- ■ floods

Damage
- 🏚 houses destroyed
- 🏠 houses affected
- ✈ travel disrupted
- 🏃 evacuations

Alerts
- 🌬 wind
- 〰 coastal event

Climate change and increasing flood risk

More storms?

Warmer ocean-surface temperatures and higher sea levels are expected to make hurricanes (cyclones) more intense, with stronger winds (2–11% stronger) and more rain (increasing by about 20%).

- In the North Atlantic, the average number of hurricanes has increased from 6 to 8 per year. This correlates with increased North Atlantic Ocean surface temperatures.
- However, recent predictions indicate that in some areas – as warmer oceans lead to more powerful, intense hurricanes – fewer storms will actually develop.

More flooding

Climate change is likely to increase the risk of coastal flooding in low-lying areas. The IPCC has predicted that, by 2100, hundreds of millions of people will have to abandon coastal zones because of rising sea levels. And, as sea levels rise, storm surges will become higher.

See page 137 of the student book on how countries can prepare for the future by adaptation and mitigation.

Ten-second summary

- Storm surges can cause severe coastal flooding with major impacts in both developing and developed countries.
- Local factors increase the flood risk on low-lying coasts.
- Climate change may increase the coastal flood risk.

 Over to you

Use this section and pages 134–136 of the student book to create a table of the economic and social consequences of coastal flooding in developed and developing countries.

You need to know:

- that hard engineering structures alter physical processes and systems
- how soft engineering works with physical processes and systems
- that sustainable management is designed to cope with future threats but can lead to conflicts.

Big idea

There are different approaches to managing the risks associated with coastal erosion and flooding.

Can coastal erosion be stopped?

Up to a point, but it's expensive and often controversial. Until the 1990s, it was usual to tackle coastal erosion using **hard-engineering** structures, but now **soft-engineering** techniques are more popular. There are advantages and disadvantages to both (Figures 1, 2 and 3 (on page 78)).

Hard engineering

This involves building structures along the coast (usually at the base of a cliff or on a beach), which alter physical processes and systems.

Type of structure	Advantages	Disadvantages	Cost
Groynes	• Work with natural processes (longshore drift) to build up the beach, increasing tourist potential. • Not too expensive.	• Interrupt longshore drift and starve beaches along the coast of sediment. • Often lead to increased erosion elsewhere.	£5000–10 000 each (at 200-metre intervals)
Sea walls	• Effective prevention of erosion. • Often have a promenade on top.	• Reflect wave energy, rather than absorbing it. • Very expensive to build and maintain.	£6000 a metre
Rip rap (rock armour)	• Relatively cheap and easy to construct and maintain.	• Rocks used are usually from elsewhere so don't fit in with local geology. • Can be very intrusive.	£100 000–300 000 for 100 metres
Revetments	• Relatively inexpensive to build.	• Intrusive and unnatural looking. • Can need high levels of maintenance.	Up to £4500 a metre
Offshore breakwater	• Effective permeable barrier.	• Visually unappealing. • Potential navigation hazard.	Similar to rock armour, depending on materials used

Figure 1 *Hard engineering – advantages and disadvantages*

Soft engineering

These techniques are designed to work with natural processes in order to manage (but not necessarily prevent) erosion (Figure 2). They can also be used to manage changes in sea level, e.g. allowing low-lying coastal areas to flood (creating marshes).

Method	Advantages	Disadvantages	Cost
Beach nourishment	• Relatively cheap and easy to maintain. • Looks natural and increases tourist potential.	• Needs constant maintenance, due to erosion and longshore drift.	£300 000 for 100 metres.
Cliff re-grading and drainage	• Re-grading can work on clay or loose rock, where other methods won't work. • Drainage is cost-effective.	• Re-grading effectively causes the cliff to retreat. • Drained cliffs can dry out and collapse.	Details unavailable.
Dune stabilisation	• Replanting marram grass maintains a natural coastal environment and wildlife habitats. • Relatively cheap and sustainable.	• Time-consuming to plant marram grass.	£200–2000 for 100 metres.
Marsh creation (a form of managed retreat)	• Relatively cheap. • Creates a natural defence, providing a buffer to powerful waves.	• Agricultural land is lost. • Farmers or landowners need to be compensated.	Cost varies, depending on the size of the area left to the sea.

Figure 2 *Soft engineering – advantages and disadvantages*

Hard engineering and Holderness

The Holderness coast is 85 km long but only 11 km are protected. Hard-engineering methods used in some places have led to problems elsewhere (Figure 3).

> **Hornsea**
> **Defences** – Sea walls, groynes, rock armour.
>
> **Impact** – Groynes trap sediment and maintain the beach at Hornsea. Mappleton has been starved of sediment and wave attack has rapidly eroded the cliffs.

> **Mappleton**
> **Defences** – Two rock groynes aim to prevent the removal of the beach by longshore drift. Rock armour is also used.
>
> **Impact** – At Cowden, 3 km south of Mappleton, sediment starvation caused increased cliff erosion.

> **Withernsea**
> **Defences** – A curved wall replaced the old sea wall in the 1990s following a cost-benefit analysis.
>
> **Impact** – Waves are noisier (breaking against the wall); the promenade is smaller.

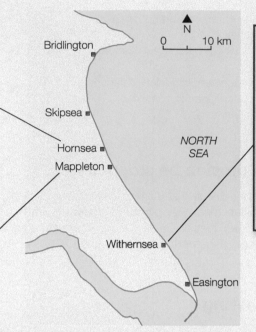

Figure 3 *Some of the impacts of hard engineering*

Cost-benefit analysis (CBA)

A cost-benefit analysis is carried out before a coastal management project is given the go-ahead. Costs are forecast and then compared with the expected benefits. Costs and benefits are of two types:

- **Tangible** – where costs and benefits are known and can be given a monetary value (e.g. building costs).
- **Intangible** – where costs may be difficult to assess but are important (e.g. the visual impact of a revetment).

A project where costs exceed benefits is unlikely to be given permission to go ahead.

 Ten-second summary

- Hard engineering involves building structures along the coast that alter physical processes and systems.
- The use of hard-engineering methods to protect one place can lead to problems elsewhere.
- Soft-engineering attempts to work with natural processes in the coastal system.
- Sustainable management can be used to deal with the threats of rising sea levels and increased storm surges.

Managing future threats

Sustainable management

The Mahanadi Delta in the Indian state of Odisha is prone to cyclone disasters.

- 50 years ago, coastal villages in Odisha had an average width of 5.1 km of mangroves protecting them. Today, the average is 1.2 km.
- In 1999, during 'super cyclone' Kalina, villages that still had 4 or more kilometres of mangroves, had no deaths. Where the mangrove belt was under 3 km wide, deaths rose sharply.

Villagers are now being helped to reverse mangrove destruction by planting mangroves along the coastline and on the banks of all tidal rivers along Odisha's coast.

For an example of managing future threats in south-west Bangladesh that has led to conflict, see pages 140–141 of the student book.

 Over to you

Explain, using diagrams, how hard-engineering methods alter physical systems and processes.

You need to know:

- that Integrated Coastal Zone Management is used increasingly to manage coasts
- that many factors are involved in making decisions on coastal management
- that coastal management can lead to conflict, with winners and losers.

Big idea

Coastlines are increasingly managed by holistic Integrated Coastal Zone Management.

Integrated Coastal Zone Management (ICZM)

ICZM is a strategy that involves managing whole sections of coast. This is because human actions in one place affect other places along the coast and because sediment moves in **sediment (littoral) cells**.

- ICZM brings together those involved in the development, management and use of the coast.
- It aims to establish sustainable levels of economic and social activity, resolve conflicts, and protect the coastal environment.

Odisha, India

Odisha's coastal zone (Figure 1) has a wide range of flora and fauna, and is rich in mineral deposits. Cultural sites dot the coast and fishing employs large numbers.

The coastal zone is under stress from:

- rapid industrialisation
- fishing and **aquaculture**
- tourism
- mining; oil and natural gas production
- erosion; rising sea levels and an increase in the frequency and intensity of cyclones.

An ICZM project has been implemented to manage the coast and its resources in a **sustainable** way. Organisations that have been consulted include: Central (Federal) government, state and local government departments, plus stakeholders in the local economy. Public consultations have also been held about issues including:

- controlling coastal erosion
- developing eco-tourism
- planting/replanting mangroves.

▲ **Figure 1** Odisha, India

Holderness

The East Riding of Yorkshire Council developed an ICZM, which was used to develop the Flamborough Head to Gibraltar Point **Shoreline Management Plan (SMP)**, published in 2011. Flamborough Head and Gibraltar Point are the northern and southern limits of a major sediment cell on England's east coast.

- The SMP sets out the policy for managing the coastline and responding to coastal erosion (and flood risks) over the next 100 years.
- The Council worked with several players and stakeholders in developing the SMP, including: national and local government agencies, and environmental and economic stakeholders.

Options for managing the coast

SMPs are recommended by the Department for Environment, Food and Rural Affairs for all sections of the coastline in England and Wales. Four options are considered for any stretch of coastline:

- hold the line
- advance the line
- managed retreat/strategic realignment
- do nothing/no active intervention.

The plan for Holderness

In order to decide what and where to protect, a **cost-benefit analysis** (CBA) and an **Environmental Impact Assessment** (EIA) are carried out. For areas on the Holderness coast (Figure 2), the CBA identified whether benefits outweighed costs, or vice versa.

An EIA decides whether environmental quality will improve, or worsen, as a result of management options. The decision under the SMP is to 'hold the line' for current defences at Dimlington and Easington gas terminals. An EIA recommended the protection scheme of a rock revetment made up of granite boulders.

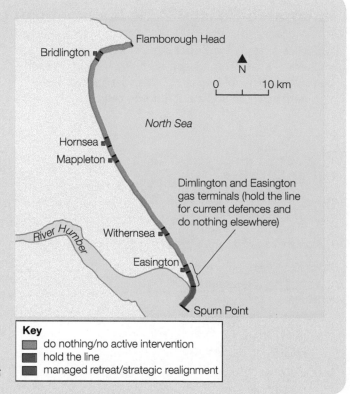

▶ **Figure 2** *Coastal management options for Holderness up to 2025*

Key
- do nothing/no active intervention
- hold the line
- managed retreat/strategic realignment

Wider issues

Decisions about whether to defend the coast are complex judgements, based on a range of factors (Figure 3).

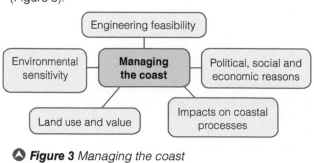

▲ **Figure 3** *Managing the coast*

Winners and losers

Many of the world's coastal zones face threats from a range of factors. Different players are involved and, as decisions are made about managing the issues, some people win, whilst others lose out.

Ten-second summary

- ICZM means that complete sections of the coast are managed as whole.
- Options for protecting the coast are complex judgements based on a range of factors.
- CBA and EIAs help to make the decision about what, and where, to protect.
- Decisions about coastal defence mean that some people lose out.

Over to you

Make a large copy of Figure 3. Use Figure 4 on page 145 of the student book to complete the details for Holderness.

Chapter 4
Globalisation

What do you have to know?

This chapter studies how globalisation is accelerating, resulting in changing opportunities, inequalities within and between countries as shifts in patterns of wealth occur, and cultural impacts as flows of ideas, people and goods take place.

The specification is framed around three enquiry questions:

1 What are the causes of globalisation and why has it accelerated in recent decades?
2 What are the impacts of globalisation for countries, different groups of people and cultures and the physical environment?
3 What are the consequences of globalisation for global development and the physical environment and how should different players respond to its challenges?

The table below should help you.

- Get to know the key ideas. They are important because 12-mark questions will be based on these.
- Copy the table and complete the key words and phrases by looking at Topic 3 in the specification. Section 3.1 has been done for you.

Key idea	Key words and phrases you need to know
3.1 Globalisation is a long-standing process which has accelerated because of rapid developments in transport, communications and businesses.	globalisation, global connections, interdependence and flows, 'shrinking world', communication, social networking, time-space compression
3.2 Political and economic decision making are important factors in the acceleration of globalisation.	
3.3 Globalisation has affected some places and organisations more than others.	
3.4 The global shift has created winners and losers for people and the physical environment.	
3.5 The scale and pace of economic migration has increased as the world has become more interconnected, creating consequences for people and the physical environment.	
3.6 The emergence of a global culture, based on western ideas, consumption, and attitudes towards the physical environment, is one outcome of globalisation.	
3.7 Globalisation has led to dramatic increases in development for some countries, but also widening development gap extremities and disparities in environmental quality.	
3.8 Social, political and environmental tensions have resulted from the rapidity of global change caused by globalisation.	
3.9 Ethical and environmental concerns about unsustainability have led to increased localism and awareness of the impacts of a consumer society.	

You need to know:
- how containerisation has accelerated
- that the Internet has increased global connections
- about global economic shifts.

Big idea
Container shipments reflect the shift in economic power from Europe to Asia.

Containerised shipments

The world's largest trade route is between Asia and Europe (Figure 1). The huge distances between ports make the size of ships and the number of containers they carry important in order to reduce costs.

- In 1990, the average container ship held just 4000 containers, and there were many shipping companies.
- Now, fewer but larger shipping companies dominate global trade and the largest ships carry 20 000 containers.

Containerised shipments have shifted the balance of economic power from Europe towards Asia.

- Products are made by European- or US-owned companies in Asia and transported to Europe.
- Relocating (or **out-sourcing**) production to Asia exploits cheaper labour costs.
- The ships, which bring high value goods to Europe, return to Asia carrying low-value waste.

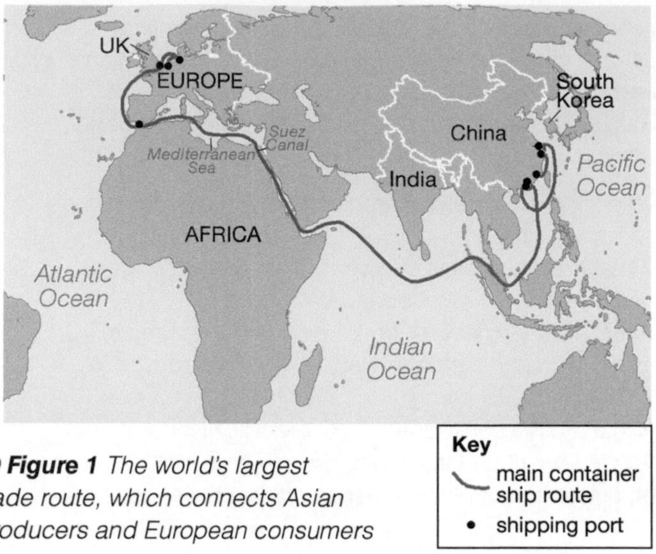

Figure 1 *The world's largest trade route, which connects Asian producers and European consumers*

Key
- main container ship route
- shipping port

A world without borders

On some container ships, many of the products will have been ordered through Amazon.

- Amazon is a product of the Internet age and, as an **e-tailer** (an electronic online retailer), it has re-shaped the retail industry.
- Quick delivery times allow Amazon to take advantage of global connections to reduce the costs of storing items in warehouses.
- Operating in most countries (Figure 2), Amazon now works in a world without borders. As a result, national governments find it hard to keep track of its sales in each country.
- Items purchased through Amazon are often cheaper because of lower operating costs and bulk buying (**economies of scale**). Economies of scale and a race to deliver ever-cheaper goods – '**the race to the bottom**' – mean that people now buy and throw away more.

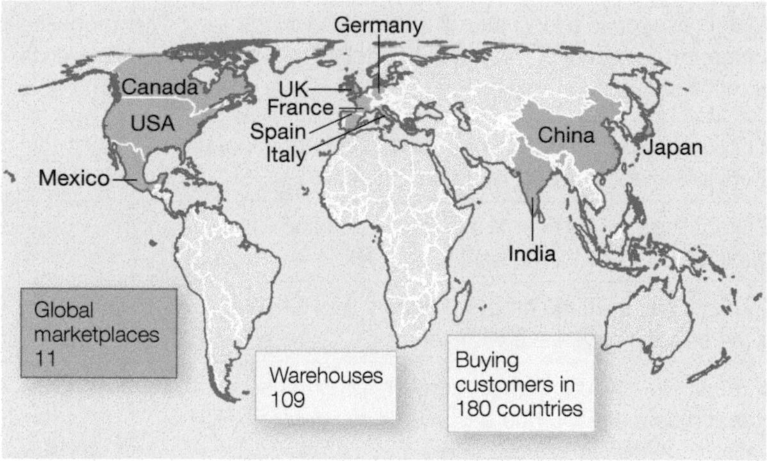

Global marketplaces 11

Warehouses 109

Buying customers in 180 countries

Figure 2 *Amazon's global business in 2015*

Ten-second summary
- Containerisation and the Internet have increased global connections and flows.
- Containerised shipments have shifted the balance of economic power from Europe towards Asia.
- Global connections encourage the development of a borderless world.

Over to you
From memory, define the following terms:
- **a** containerisation
- **b** economies of scale
- **c** out-sourcing.

You need to know:
- what globalisation is
- how globalisation has accelerated.

Big idea

Globalisation is not a new process, but has accelerated due to rapid developments in transport, communications and business.

What is globalisation?

Globalisation is no longer just about the movement of raw materials (**commodities**) or goods. It is the process by which people, culture, finance, goods and information transfer between countries with few barriers.

- Physical distances haven't changed. But developments in technology have massively reduced the time it takes to trade and communicate globally. Developments in transport have also reduced actual travel times.
- This process is called **time-space compression**, and it has led people to refer to a '**shrinking world**' (Figure 1).

How globalisation works

- Those who make decisions to invest and manufacture overseas come mainly from North America, Japan and Europe, as well as oil-rich billionaires.
- China, India and South-East Asia have become manufacturers for the world.
- India also provides financial and IT support services for HICs.

Outsourcing and relocation processes often change to where costs are even lower (such as Vietnam or Bangladesh). Meanwhile, much of sub-Saharan Africa remains detached and isolated, with little economic influence.

FALLING COSTS OF COMMUNICATIONS

1500–1840
sailing ships
averaged 10 mph

1850–1930
steam trains
averaged 65 mph

1950s
propeller aircraft
300–400 mph

1960s
jet passenger aircraft
500–700 mph

1990s–today
cyberspace
information in seconds

REDUCING TIME LAPSE OF INFORMATION TRANSMISSION
Morse code → telephone → satellite, fibre optics/internet and broadband

▶ **Figure 1** *A shrinking world – time-space compression affects some places more than others, depending on the connections*

The processes and impacts of globalisation

Financial

- Global capitalism is spread by large TNCs.
- Cheaper labour in developing economies helps supply wealthier nations with goods.
- Trillions of dollars are exchanged globally by electronic means every day.

Political

- Some TNCs seek to influence how people think.
- TNCs and international political organisations can influence national governments.
- Many trade barriers have been reduced or removed.

People and migration

- People with management, finance and IT skills move to meet demand.
- Economic migrant labour flows to areas with higher incomes.

Communication and information

- Lower transport costs allow increasing long-distance tourism.
- Cheaper global phone networks and fast fibre-optic connection are available.
- Exchanges of people, information and ideas become commonplace.

 Ten-second summary

- Globalisation is the process by which people, culture, finance, goods and information transfer between countries.
- Time-space compression has led to a 'shrinking world'.
- The processes and impacts of globalisation involve finance, politics, populations and communication.

Over to you

In no more than six sentences, summarise how the following have led to a 'shrinking world':

a developments in transport and trade

b rapid development in ICT and mobile communication.

You need to know:

- about the role of political and economic organisations, and national governments, in globalisation
- how special economic zones, subsidies and foreign direct investment have contributed to the spread of globalisation.

Big idea

Political and economic decision-making are important factors in the acceleration of globalisation.

Global players

Increasingly, decisions that affect people locally are being made by global organisations. These are **players** in globalisation.

Three global organisations were established in 1945, following the Second World War, as key players to promote economic development and to restore and maintain financial stability. They remain fundamental to global decision-making today.

The World Bank

Its purpose is to use bank deposits placed by the world's wealthiest countries to provide loans for development.

- Recipient countries have to agree to certain conditions concerning repayment and economic growth.
- It also focuses on natural disasters and humanitarian emergencies.

The International Monetary Fund (IMF)

Its purpose is to maintain international financial stability.

- In return for loans, it tries to force countries to privatise (or sell off) government assets in order to increase the size of the private sector and generate wealth.
- Many believe that this policy has forced poorer countries to sell off their assets to wealthy TNCs.
- It also exists to stabilise currencies in order to maintain economic growth.

The World Trade Organisation (WTO)

Its purpose is to promote free flow of trade to prompt economic growth, especially in the poorest countries.

- The WTO believes in Free Trade, advocates removing **barriers** ('**trade liberalisation**') and seeks to encourage all trade between countries free of **tariffs**, **quotas** or **restrictions**.

International trading blocs

Increasingly, countries are grouped together as members of **trading blocs**, (e.g. EU and NAFTA, Figure 1) to promote free trade. They support trade for their members by:

- removing tariffs between member states
- placing tariffs on imports from non-member states to protect member states' industries.

▶ **Figure 1** *Four of the world's trading blocs*

Key

- **NAFTA:** USA, Canada, Mexico
- **ASEAN:** Brunei, Cambodia, Indonesia, Laos, Malaysia, Myanmar, Philippines, Singapore, Thailand, Vietnam
- **APEC:** Australia, Brunei, Canada, Chile, China, Indonesia, Japan, South Korea, Malaysia, Mexico, New Zealand, New Guinea, Peru, Philippines, Russia, Singapore, Taiwan, Thailand, USA, Vietnam
- **Cairns Group:** Argentina, Australia, Brazil, Canada, Chile, Colombia, Costa Rica, Guatemala, Indonesia, Malaysia, New Zealand, Pakistan, Paraguay, Peru, Philippines, South Africa, Thailand, Uruguay, Vietnam

Individual national governments

The UK

Just as important to the globalisation process is the willingness of individual national governments to promote international strategies for growth. In the 1980s, the UK Conservative government developed two strategies:

- **Tax breaks (subsidies)** – These have encouraged a number of large overseas financial institutions to relocate to London (e.g. to Canary Wharf).
- **Grants and subsidies** – These have encouraged foreign companies to locate new manufacturing plants in the UK (e.g. Toyota in Burnaston, Derbyshire).

China

After decades of economic and political isolation, the Chinese government declared an '**open door**' policy to international business in 1978.

- Companies from Europe and the USA quickly saw the advantages of out-sourcing and relocating into one of southern China's four '**special economic zones**', later known as '**Export Processing Zones**'.
- These zones offered tax incentives and huge pools of cheap labour. Since then, China's economy has grown rapidly.

Investment like this from overseas companies is known as **Foreign Direct Investment (FDI)**.

How have flows of FDI changed?

- China is still the world's largest recipient of FDI.
- Now countries such as India, Brazil, Russia and South Africa (BRICS countries), also control flows of FDI.
- The BRICs invest heavily in the USA, EU, sub-Saharan Africa (Figure 2) and South America.
- Investment flows from the BRICS to other countries now account for over 10% of the global total.

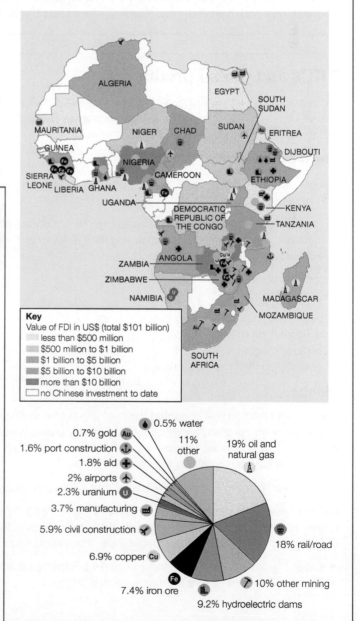

▶ *Figure 2* FDI flows from China into Africa

See pages 156–157 of the student book for case studies of how the WTO affects trade for Ghana, Vietnam and Guatemala.

Ten-second summary

- The IMF, World Bank and WTO promote free trade policies and FDI.
- In trading blocs, tariffs are removed between member states, which can help countries to develop.
- National governments have policies for promoting economic growth, such as grants and subsidies.
- China's 'open door' policy has encouraged FDI into China.

Over to you

Write down:

a three global organisations that promote economic development

b two ways those organisations have contributed to globalisation

c one consequence of FDI.

You need to know:

- how TNCs contribute to the spread of globalisation
- how TNCs take advantage of economic liberalisation.

Big idea

Globalisation has affected some places and organisations more than others.

Transnational corporations: corporate colonialism?

Transnational corporations (TNCs) are vital players in globalisation. In 2016, there were over 60 000 TNCs. The top 200:

- employed just 1% of the global workforce, but
- accounted for 25% of the world's economic activity.

However, a new world order is emerging.

- In 2006, six of the top ten TNCs were based in the USA.
- By 2015, only two of the top ten were American, and three Chinese companies had overtaken Ford, General Motors and ConocoPhillips.

Rank	Company	Country of origin	Turnover 2015 (US$ billion)	Profit 2015 (US$ billion)	Revenue compared to the GDP of whole countries
1	Walmart Stores	USA	485.6	16.3	More than Sweden
2	Sinopec	China	446.8	5.1	More than Norway
3	Royal Dutch Shell	Netherlands	431.3	14.8	More than Norway
4	PetroChina (CNPC)	China	428.6	16.3	More than Norway
5	Exxon Mobil	USA	382.6	32.5	Equal to Austria

Figure 1 Fortune *magazine's top five TNCs in 2015*

TNCs and global production

TNCs are vital to the spread of globalisation because their expansion involves the free flow of capital, labour, goods and services. Three factors have led to this – motive, means and mobility.

Motive

Under capitalism, there is one motive – profit. Companies become dominant by:

- controlling and minimising their costs (e.g. raw materials, labour)
- increasing their revenues by expanding their markets and merging with or taking over their competitors.

Corporations manage this through:

- achieving economies of scale
- developing new markets
- horizontal integration
- vertical integration
- diversifying their product range.

Means

Banking and the free flow of **capital** (money) around the world are the mechanisms for company growth. Global, relatively unrestricted flows of finance connect banks, businesses and countries in complex webs. These flows between countries are variable:

- from year to year   • over the medium term   • over the longer term.

A kind of **reverse colonialism** has happened. Hong Kong, Singapore, China, South Korea, Malaysia, India and Brazil are now established HICs or MICs and are all net providers of overseas investment (Figure 2).

- In 2015, there were over 800 Indian-owned businesses in the UK.
- China is investing heavily in infrastructure in Africa.

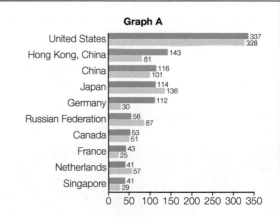

Graph A

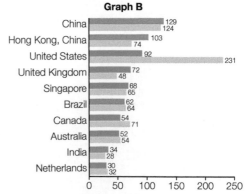

Graph B

Key
2013   2014
Developed economies
Developing and newly industrialising economies

Figure 2 *The top ten countries for FDI in 2013 and 2014: Graph* **A** *shows the source countries and Graph* **B** *shows the recipients. The figures at the end of each bar are in US$ billion*

Mobility

Mobility has been fundamental to the spread of globalisation. This includes:

- faster and cheaper transport
- rapid communication systems
- new production technology, such as **just in time**
- global production networks run from company headquarters, almost always in HICs.

However, mobility flows mean that production and sources of materials can be flexible, which can threaten jobs if companies move elsewhere.

 For more detail on motive, means and mobility, see pages 158–159 of the student book.

New economy – new heroes?

Disney is typical of the **new economy**, where ideas are as important as goods. Disney's ideas originate within the USA, but its merchandise is produced overseas. It operates a just-in-time system for merchandise production, which has both benefits and problems.

- It uses overseas manufacturers – called **off-shoring** or **outsourcing** – and demands quick delivery times. This avoids having to operate its own expensive production lines.
- However, outsourcing is not problem-free. Overseas workers often earn low wages (e.g. in Vietnam). Overseas factories have also used toxic substances that are banned in the USA.

Cultural globalisation and glocalisation

Disney owns 40 Spanish-speaking radio stations, together with foreign language television channels and a Chinese radio station in Hong Kong. It also supplies reading materials to help teach English in Chinese and African schools. It was one of the first global TNCs to **glocalise**, by tailoring its products to specific market areas. (Figure 4).

- *The Hunchback of Notre Dame* was launched to rebrand Disneyland Paris.
- *The Lion King* was aimed at African markets.
- *Mulan* marked a decision to enter the Chinese market.

- Over 250 satellite and cable TV companies
- 6 film/TV production and distribution companies, including Pixar and Lucas film
- 12 publishing companies and 15 magazines/newspapers
- 728 shops worldwide
- 5 record labels and music publishing companies
- 14 theme parks and resorts, and a cruise line

 Figure 3 *Some of the many global activities Disney is involved in*

Figure 4 *Disneyland Hong Kong, another example of Disney seeping into everyday lives and influencing local government, which paid US$1 billion towards this leisure location*

 Ten-second summary

- The expansion of markets and free flow of capital enable TNCs to grow.
- Mobility flows mean that production and sources of materials can be flexible.
- Offshoring/outsourcing involves using overseas manufacturers.
- Products are often tailored to specific market needs.

Over to you

Make a list of the ways in which TNCs contribute to the spread of globalisation.

You need to know:

- how degrees of globalisation can be measured using the KOF Index and the A T Kearney Index
- the physical, political, economic and environmental reasons why some locations remain largely 'switched off' from globalisation.

Big idea

Globalisation has meant that some places are more 'switched off' than others.

A two-speed world

Globalisation is contentious. Some believe it has created a two-speed world.

- Those in favour of globalisation claim that increased connectedness (becoming '**switched on**') improves many countries' economic development.
- Those against globalisation argue that it leads to corrupt practices, and that some countries and people are left behind ('**switched off**').

Degrees of globalisation

As countries become more globalised, certain characteristics (**indicators**) change. These include:

- **flows** (e.g. trade, migration)
- **technologies** (e.g. Internet usage, telecommunications)
- **movements** (e.g. air traffic)
- **media** (e.g. advertising, music).

These indicators improve connections, which are the basis of globalisation.

How degrees of globalisation vary

The KOF and Kearney Indexes measure how globalised a country has become.

Measuring globalisation 1 – the KOF Index

The **KOF Index** score for each country is calculated using specific interactions.

- **Economic globalisation** – e.g. cross-border transactions and the volume of FDI.
- **Social globalisation** – e.g. information flows, and the presence of McDonald's as an indicator of 'global affinity'.
- **Political globalisation** – e.g. the number of foreign embassies and the country's membership of international organisations.

Thirteen of the top 15 most globalised countries are European (Figure 1), which contrasts with tables of GDP (where the USA leads) or manufacturing output (where China leads). The KOF Index measures *international* interactions. China and the USA have large domestic economic markets, so *internal* connections are important. But these do not count in the KOF values.

Rank	The 15 most globalised countries and their scores in 2015	Rank	The 15 least globalised countries and their scores in 2015
1	Ireland 91.3	177	Myanmar 33.01
2	Netherlands 91.24	178	Sao Tome and Principe 32.86
3	Belgium 91	179	French Polynesia 32.82
4	Austria 90.24	180	Burundi 32.26
5	Singapore 87.49	181	Tonga 31.89
6	Sweden 86.59	182	Sudan 31.54
7	Denmark 86.3	183	Comoros 31.15
8	Portugal 86.29	184	Afghanistan 30.62
9	Switzerland 86.04	185	Bhutan 29.3
10	Finland 85.64	186	Equatorial Guinea 27.49
11	Hungary 85.49	187	Eritrea 27.13
12	Canada 85.03	188	Laos 26.91
13	Czech Republic 84.1	189	Kiribati 26.00
14	Spain 83.71	190	Somalia 25.39
15	Luxembourg 83.56	191	Solomon Islands 25.26

🔺 **Figure 1** *The world's most and least globalised countries in 2015, based on the KOF Index*

Measuring globalisation 2 – the A T Kearney Index

The A T Kearney Index identifies four main indicators:

- **Political engagement** – e.g. a country's participation in international organisations and peacekeeping operations.
- **Technological connectivity** – e.g. the number of Internet users and servers.
- **Personal contact** – e.g. through telephone calls, travel and remittance payments.
- **Economic integration** – e.g. the volumes of international trade and FDI.

This index uses more holistic indicators (e.g. the number of web servers, rather than Internet communications) and also volumes of trade as well as FDI. The A T Kearney Index 'top performers' for 2015 are shown in Figure 2.

Overall ranking	Country	Economic ranking	Personal ranking	Technological ranking	Political ranking	Best scores for
1	Singapore	1	3	12	29	Trade, FDI and personal
2	Switzerland	9	1	7	23	Phone contacts
3	USA	58	40	1	41	Internet hosts and servers
4	Ireland	4	2	14	7	Economic and personal
5	Denmark	8	8	5	6	Ranked highly across the board
6	Canada	23	7	2	10	Technological then personal
7	Netherlands	21	11	6	5	Political and technological
8	Australia	18	36	3	27	Internet users
9	Austria	15	4	13	2	Political and personal
10	Sweden	19	12	9	9	Internet users and political
11	New Zealand	35	15	4	24	Secure Internet servers
12	UK	25	14	8	4	International treaties and technological

▲ **Figure 2** *The 'top performers' in 2015, according to the A T Kearney Index*

Barriers facing sub-Saharan Africa

The case studies of Zambia and Tanzania demonstrate the physical, political, economic and environmental reasons why some countries are struggling to 'switch on'.

Zambia

Zambia is the world's eighth largest producer of raw and part-processed copper. However, it is a **landlocked** country, so it relies on good political relations with its neighbours to access ports.

- The TanZam railway takes Zambia's copper to the Tanzanian coast.
- The Benguela rail link carries copper for export to the Angolan coast.

Both have been developed and upgraded using Chinese investment.

Copper remains Zambia's biggest export, though its value has fallen in recent years.

- Since 2000, privatisation and debt cancellation have reduced Zambia's debt. US$20 billion of FDI has been invested in the copper industry, so it can part-process the ore and add value to it, increasing the country's income.

See pages 164–165 of the student book to see how Zambia and Tanzania's development indicators have changed from 2004 to 2014.

Tanzania

80% of Tanzania's working population is employed in agriculture. One of its main crops is raw cotton.

- Due to global overproduction, cotton prices frequently fall. The country is then less able to pay for imported manufactured goods.
- GDP fluctuates, so Tanzania is struggling to 'switch on'.
- The **Heavily Indebted Poor Countries (HIPCs)** initiative led to the cancellation of many of Tanzania's debts. Now the income gained from growing export crops is increasing investment in schools and healthcare.
- Tanzania also has growing investment links with countries such as India and China, in order to export its farm produce and mineral resources.

Ten-second summary

- The KOF and A T Kearney Indexes measure degrees of globalisation and identify 'switched on' countries.
- Some countries face barriers to progress and are struggling to 'switch on'.

Over to you

1 List the factors that encourage some countries to become 'switched on'.
2 Draw a table to show the physical, political, economic and environmental reasons why Zambia and Tanzania have remained largely 'switched off' from globalisation.

You need to know:

- that the global economic shift has brought benefits and costs
- about the problems some deindustrialised regions face as a result of economic restructuring.

Big idea

The global shift has created winners and losers for people and the physical environment.

The shift begins

In the 1970s and 1980s, the **global shift** began – the movement of manufacturing from Europe and the USA to many Asian countries. Countries such as Japan and South Korea – then China and India – became major players in the globalised economy.

Three factors accelerated this global shift:

- Individual Asian countries allowed overseas companies access to their markets.
- TNCs sought new areas for manufacturing (e.g. China) and for outsourcing services (e.g. India).
- FDI began to flow into emerging or re-emerging Asian countries.

The global shift – India

Many TNCs have outsourced services such as call centres and software development to India. It is an attractive location because of its:

- close political links with the UK
- widespread adoption of English
- good technical universities.

The global shift – impacts in China

The global shift into China has mostly focused on manufacturing. Rapid industrialisation has been accompanied by rapid urbanisation.

Benefits of economic growth

- **Investment in infrastructure** – By 2016, China had:
  - the world's longest highway network and high-speed rail system
  - the world's fastest commercial train service (Shanghai's Maglev, Figure 1)
  - eight of the world's top 12 airports, by freight tonnage.
- **Reductions in poverty** – From 1981 to 2010, China reduced the number of people living in poverty by 680 million. From 1980 to 2016, its extreme poverty rate fell from 84% to 10%.
- **Increases in incomes** – Urban incomes have risen by 10% a year since 2005. There is a big and growing rural-urban divide (Figure 2).
- **Better education and training** – Education is free and compulsory for 6–15 year olds. In 2014, 7.2 million Chinese graduated from university, and this has helped to create a skilled workforce. Again, there is a big rural-urban divide.

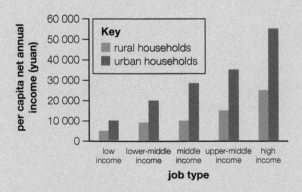

▲ *Figure 1* *Shanghai's Maglev train links the airport and the CBD*

▲ *Figure 2* *Variations in annual disposable income (in yuan) between urban and rural areas in China, by job type*

Costs of economic growth

- **The loss of productive farmland** – By 2016, over 3 million hectares of arable farmland in China had been polluted with heavy metals. Increased use of fertilisers and pesticides has led to farmland near rivers being taken out of production.
- **Increase in unplanned settlements** – The need for more urban housing has resulted in an increase in two types of **informal housing**, both illegal:
  - Villagers on the edge of cities add extra storeys to their houses, then let the space.
  - Farmland is developed for housing without permission.
- **Pollution and health problems**
  - In 2015, one US climate research organisation calculated that Chinese air pollution kills an average of 4400 people every day.
  - 70% of China's rivers and lakes are now polluted.
  - 360 million Chinese do not have access to safe drinking water.

- **Land degradation** – 40% of China's farmland is now suffering degradation. Soils are eroding or suffering from acidification caused by industrial emissions.
- **Over-exploitation of resources and resource pressure** – China needs more resources. Amazonian rainforest has been cleared in Ecuador and oil fields are under development in Venezuela – all for China's consumption.
- **Loss of biodiversity** – In 2015, WWF found that China's terrestrial vertebrates have declined by 50% since 1970. The main causes are habitat loss and the degradation of natural environments by economic development.

The global shift – impacts in the UK's industrial cities, e.g. Leicester

Some regions in HICs also face social and environmental problems as a result of the global shift.

- For example, Leicester was once dominated by the textile industry.
- However, by the 1970s, cheaper clothes were manufactured in Asia. Industries closed, causing **deindustrialisation**.

Dereliction and contamination

Many textile companies in Leicester were forced to close (Figure 3). A lot of the land was left abandoned or **derelict**. Derelict industrial land can often be contaminated.

Unemployment, depopulation and deprivation

In the 1970s and 1980s, many inner-city areas became run down and the housing was low cost.

- People on low incomes or unemployment benefit moved in. The areas became pockets of deprivation.
- In Leicester, areas of deprivation often coincide with previous industrial areas.

Many inner-city areas have gained reputations for crime. In fact, crime rates have fallen sharply since 2000. This decline may be due to regeneration and gentrification.

Figure 3 *The impact of deindustrialisation – one of Leicester's former textile mills, now semi-derelict*

 Ten-second summary

- The global shift is the movement of manufacturing from Europe and the USA to many Asian countries.
- It has brought benefits and problems to China, including reductions in poverty and pollution.
- Cities in the UK such as Leicester experienced deindustrialisation, which has brought dereliction and deprivation.

 Over to you

Draw two mind maps to explain the impacts of the global shift, one for each of:

a China
b Leicester.

You need to know:

- the causes of rapid urban growth and how this growth creates challenges
- how international migration has deepened interdependence
- how migration has costs and benefits for both host and source locations.

Big idea

Economic migration has increased with greater interconnection, creating consequences for people and the environment.

It's an urban world

Over half of the global population now lives in urban settlements.

- The numbers of **million cities** (populations over 1 million) and **mega cities** (populations over 10 million) are growing rapidly.

- **World cities** (or **hub cities**) are cities with a major influence, such as London and New York. They are where political and economic decisions are made. They attract flows of economic migrants, as well as of capital, and are the best-connected cities.

Hyper-urbanisation in New Delhi

New Delhi is experiencing **hyper-urbanisation**. It's predicted to grow by 40% between 2010 and 2020. The main causes are:

- high birth rates and low death rates (i.e. a high rate of natural increase)
- high rates of rural-to-urban migration.

Rural migrants consist of:

- the rural poor, who lack opportunities in their villages
- the rural rich, who want a better standard of living.

Some factors **pull** (or attract) people into cities. Other factors **push** people away from rural areas (Figure 1).

In New Delhi, increased FDI has created many new jobs. Its financial district has become a global finance hub (Coca-Cola and Microsoft base their Indian operations there).

However, many of New Delhi's rural migrants end up living in slums and struggling to find work.

Social challenges caused by rapid growth

- Governments are challenged to provide services such as housing and education.
- Private companies often provide services. They target high earners first. The wealthier areas have piped safe water, while poorer areas don't.
- Shanty towns occur in LICs and MICs.
- In wealthy cities, the number of homeless people rises as accommodation becomes more unaffordable.

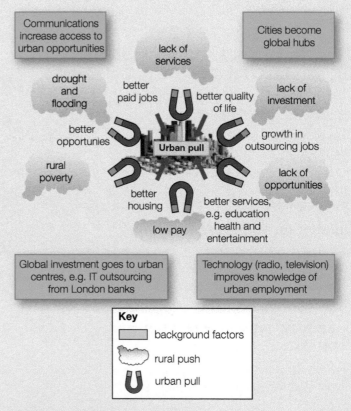

▲ **Figure 1** *Urban growth factors*

Environmental challenges caused by rapid growth

- Air quality – New Delhi is judged to have the worst (out of 1600 cities worldwide). The number of vehicles there is predicted to rise from 4.7 million in 2010 to 26 million by 2025. Air pollution is India's fifth largest killer.
- Environmental problems also include sewage pollution and chemical dumping.

International migration

Globalisation means many people move around the world more freely, increasing **interdependence** between regions.

Elite migrants – London

London is a major world city. Flows of skilled, wealthy migrants – **elite migrants** – have moved there.

Overseas buyers have invested in a number of London properties:

- Qatari Investments has bought into The Shard (Figure 2) and Harrods.
- Between 2004 and 2014, a third of all purchases of residential property in London by overseas buyers went to wealthy Russians.
- This high demand has led to property price inflation.
- Much London housing has become too expensive for many local people.

Low-wage migrants – UAE and Qatar

The United Arab Emirates (UAE) and Qatar each has labour shortages.

- They recruit manual workers from overseas (Figure 3).
- Migrant workers now make up 90% of UAE's workforce (including 1.75 million Indians and 1.25 million Pakistanis).
- Movements on this scale are referred to as **mass low-wage economic migrations**.

The costs and benefits of migration

Voluntary international migrations are mostly economic. There are gains and losses for the host and source locations (Figure 4).

The host location...	The source location...
receives skilled foreign workers	experiences reduced unemployment, as people emigrate to find jobs elsewhere
gains lower-skilled workers for jobs that are difficult to fill	
can sustain the lifestyle of its middle classes by gaining workers to provide childcare, cleaning and elderly care	loses its most skilled and dynamic workers (the brain-drain)
	earns remittance payments, sent home by overseas migrants
can often balance an ageing population with young adults	suffers an imbalanced population because many young people migrate, leaving a dependent population.
can experience pressure on housing, healthcare and school places.	

 Figure 4 *The impacts of international migration on host and source locations*

The politics of migration

Migration is a political issue too. In 2015, there was a massive increase in the number of illegal migrants entering the EU. Several European countries erected fencing along their borders to try to control the flow.

Figure 2 *The Shard in London is owned by Qatari Investments*

Figure 3 *Low-wage migrant workers helping to build new facilities in Qatar for the 2022 football World Cup*

Ten-second summary

- Rapid urban growth creates challenges such as pollution and the need for adequate services.
- International migrants include elite migrants and low-wage migrants.
- Migration has costs and benefits for host and source locations.

Over to you

Write down your top three facts to remember about:

a hyper-urbanisation in New Delhi
b elite migrants in London
c low-wage migrants in UAE.

You need to know:

- how cultural diffusion occurs as a result of globalisation
- how cultural erosion has resulted in changes to the environment
- why there is opposition to globalisation from some groups.

Big idea

The emergence of a global culture is one outcome of globalisation.

Cuba

In Cuba, strict communist controls have been relaxed recently. This has allowed **free enterprise** businesses to be set up for the first time in decades.

- Cuba's new spirit of openness has led to improved relations with the USA.
- However, social and economic changes – which accompany political relaxation – are leading to growing inequality.

Cuba is changing in other ways:

- The influx of international tourists, and the spread of satellite TV and the Internet, is broadening Cubans'

knowledge of the rest of the world. This knowledge challenges Cuba's traditions and values – globalisation is diluting Cuban culture.

- In some locations, **cultural erosion** (e.g. the loss of language and traditional food) has also resulted in changes to the environment. Beach resorts have changed Cuba's coastline. Its coral reefs are now threatened by increased tourist activity.
- The processes by which Western values have moved to Cuba are known as **cultural diffusion**. It is a symbol of globalisation. Faster connections enable the spread of attitudes around the world and traditions are lost to new trends.

Cultural landscapes and diversity

Current global changes are affecting the ways in which people see the world.

- **Glocal** cultures develop where global processes exist at a local level (e.g. Figure 1).
- Urban environments (like Leicester's Golden Mile) have been transformed by decades of inward migration.
- **Ethnic enclaves** gain their own identity. Street furniture, road names and cuisine add to the city's multicultural character.

⬆ **Figure 1** *Diwali celebrations light up the Golden Mile in Belgrave Ward, Leicester*

Cultural diffusion and the media

The ownership of the global film, broadcasting and music industries has become more concentrated into the hands of large media TNCs. Therefore, the use of an increasingly common vocabulary is starting to erode cultural diversity. This is known as the **global homogenisation** of culture – with everything becoming the same (Figure 2).

Newspapers and magazines

- Australia: 101 newspapers.
- UK: Four newspapers, e.g *The Times* and *The Sun*.
- USA: *The New York Post*, *The Wall Street Journal* and other regional papers.
- Russia: 33% share in Russia's leading financial newspaper.

Television

Studios, networks and satellite channels in countries such as France, India, Isreal and the USA. These include BSkyB (UK), STAR TV (Asia) and Twentieth Century Fox.

⬆ **Figure 2** *Some companies worldwide owned by News Corp. This spread of global culture can impact on people's political as well as cultural thinking, e.g. every winning political party in UK General Elections since 1979 has been promoted by* The Sun

Cultural exchanges

Information technology and digital communications spread ideas and products faster than ever. What we see, read or listen to is increasingly provided by a small group of huge companies. Five companies now own 90% of the global music market.

In some ways, globalisation is the 21st-century term for 'cultural imperialism'. Previous generations called it 'Americanisation', 'Westernisation' or 'Modernisation'. As the global economy draws people together, brands such as Coca-Cola become globally famous. Every year, Forbes (a US company) calculates the brand value of 500 top companies. Its top ten for 2015 are shown in Figure 3.

Ranking	Brand	Type of company
1	Apple	Technology
2	Microsoft	Technology
3	Google	Technology
4	Coca-Cola	Beverages
5	IBM	Technology
6	McDonald's	Restaurants
7	Samsung	Technology
8	Toyota	Automotive
9	General Electric	Diversified
10	Facebook	Technology

Figure 3 The Forbes top ten brands in 2015

Changing values

Globalisation can also challenge values, sometimes in a transformative and beneficial way.

- In 2012, author James Palmer wrote: 'Disabled people in modern China are still stigmatised, marginalised and abused.'
- That same year, China came top of the medals table in London's Paralympic Games.
- The chance to train and compete on the global stage has helped those with disabilities to gain a more equal status.
- Even so, there is a long way to go before equality is achieved.

Increasingly, Western countries are adopting more tolerant policies on other ethical issues, such as gay rights. However, there is still some way to go in places such as Russia, the Middle East and Africa.

Growing resistance

The spread of global culture has been counterbalanced by concerns about its impacts.

- Concerns include the perceived exploitation of environments and people.
- This has led to opposition by anti-globalisation groups and environmental pressure groups.
- They reject globalised cultures and the practices of many TNCs in avoiding tax payments.

Two examples of recent national resistance to a 'global culture' are:

- In **Iran** in the early 2000s, Barbie dolls were confiscated from toy stores because the government denounced Barbie's un-Islamic image. Since then, the government has liberalised its position.
- Until the early 2000s, the **French** government limited how much foreign culture could be broadcast. However, Internet downloading of music and films placed this in dispute. Since 2007, the French government has been more accepting of globalisation.

Ten-second summary

- Closer and faster connections enable cultural diffusion.
- TNCs have a large influence in the spread of culture.
- Globalisation can lead to increasing opportunities for disadvantaged groups.
- The spread of globalisation is opposed by some groups.

Over to you

Learn the contents of this section. Tomorrow, use a voice recorder to record what you can remember. Use the following prompts:

a cultural erosion and diffusion
b cultural exchanges
c changing values
d growing resistance.

Then use this section to check how much you remembered.

You need to know:

- how economic, social and environmental development indices differ
- how widening inequality suggests that globalisation has created winners and losers
- how countries have made differing progress in economic development and environmental management.

Big idea

Globalisation has led to dramatic increases in development for some countries. It has also widened the development gap and led to differences in environmental quality.

Richer or poorer?

Globalisation has created an explosion in global trade. But this process has led to greater inequalities.

- The GDP of most Asian countries has accelerated rapidly.
- China's richest 1% own one-third of China's property and industrial wealth.
- Every global region increased its GDP per capita from 1980 to 2012 (Figure 1).
- However, the **development gap** (the difference between the wealthiest and poorest nations) actually widened.

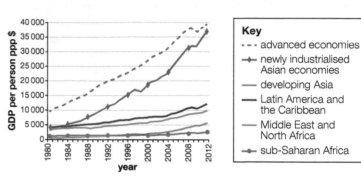

🔺 **Figure 1** *The changing GDP per person (in PPP$) in different economic world regions between 1980 and 2012*

Measuring development

Economic measures

Economic measures can be used to measure progress and include **single indicators** such as:

- **GNI (Gross National Income)** – the value of goods and services earned by a country (including overseas earnings).
- **GDP (Gross Domestic Product)** – the same as GNI, but excluding foreign earnings.
- **Purchasing Power Parity (PPP)** relates average earnings to local prices and what they will buy.

Other indicators exist in **composite** form, i.e. using several sets of data.

- The most common is **economic sector balance** (the percentage contribution of primary, secondary and tertiary sectors to GNI).
- As countries develop manufacturing industries, the value of their primary sector output falls and the secondary sector rises.
- In some LICs, such as Malawi, the primary sector value is still high (30%), while the UK's is low (0.6%).

Social indicators

The **Human Development Index (HDI)** shows how far people are benefiting from economic growth (Figure 2). It uses four indicators:

- life expectancy
- education (using two indicators: literacy and the average number of years of education)
- GDP per capita (using PPP$).

The indicators are combined into a single value ranging from 0 (low) to 1 (high).

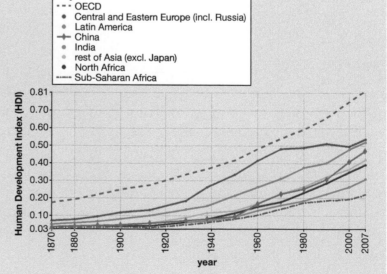

▷ **Figure 2** *UN estimates of trends in the HDI from 1870 to 2007*

Women and development

In no country are women equal to men. So the UN has developed the Gender Inequality Index (GII), which uses indicators relating to:

- **Reproductive health** – as gender inequalities decline, fertility rates and maternal mortality rates fall and the age of having a first child rises.
- **Empowerment** – women enter politics as they become empowered.
- **Education and employment** – staying in education opens up more opportunities for women.

Environmental quality indicators

Air quality is an indicator of economic development. It deteriorates as development increases. Most countries calculate air quality indices but in slightly different ways, so international comparisons are difficult.

A widening gap?

There is a gap between the economic growth rates in different global regions. But widening income inequalities also develop *within* countries. This can be measured using the **Gini index**, an index with values between 0 and 100 shown using a Lorenz curve (Figure 3).

- A low index value indicates a more equal income distribution.
- A high index value indicates unequal distribution.

Only about one-third of countries publish a Gini index.

See Section 7.7 in the student book for an explanation of how the Gini index is calculated.

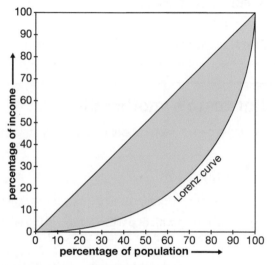

▲ **Figure 3** *The Lorenz curve, used to plot and calculate the Gini index*

China's economic and environmental inequalities

An east–west divide now exists in China.

- Incomes decline further away from the coast (in the west).
- Almost all the major cities and industrial zones are located on the coast.
- The 2010 Gini index for China as a whole was 47 and its inequality is increasing.

Chongqing is at the heart of economic developments around the Three Gorges Dam. Its filthy air causes thousands of premature deaths. China may be doing well economically, but is it at the expense of the environment?

 Ten-second summary

- Development is measured economically, socially and environmentally.
- Globalisation has led to greater inequalities, both between and within countries.
- There are winners and losers for people and the environment.

 Over to you

Cover up everything in this section except Figure 1. Use the graph to explain:

a how globalisation has contributed to regional inequalities
b the potential social and environmental consequences of the data shown.

You need to know:

- how globalisation has created culturally mixed societies and also tensions
- what attempts have been made to control the spread of globalisation
- how some groups seek to maintain their cultural identity.

Big idea

Social, political and environmental tensions have resulted from rapid global change.

London's melting pot

London has residents from every country, speaking almost every language. This growing diversity has been helped by several processes, which are linked to globalisation:

- **Open borders** – EU citizens are free to move around the EU.
- **The freedom to invest in business or transfer capital** – In the UK, any bank or individual can trade in shares without having to use the London Stock Exchange.
- **FDI** – In 2015, the UK attracted over 32 000 jobs (mostly to London) from overseas-owned companies investing in software and financial services.

Processes like those above (and others such as historical links) have led to a cultural mix.

But there's another view

Not everyone is happy about globalisation and its effects. Two big issues in relation to this are:

Immigration

International migration can cause resentment in some people within host populations.

- **Extreme political parties**, such as the Northern League in Italy, are becoming increasingly popular.
- Since 2014, streams of refugees from Syria – as well as economic migrants from many other countries – have caused tensions between Greece and other Balkan countries, and Turkey.

Trans-border water conflicts

The Mekong is one of South-East Asia's major rivers. The **Mekong River Agreement** required the governments of Cambodia, Laos, Thailand and Vietnam to all agree to any proposals for new dams before they go ahead. The impacts of building dams along the Mekong are likely to be far-reaching (Figure 1).

China
Damming the Mekong to generate hydroelectric power would encourage economic development in China's poor south.

Thailand
Only 36% of Thailand's territory is within the Mekong basin, but Thailand would like the water and electricity generated by the dams for industrial development.

Laos
90% of Laos' population depends on the Mekong for agriculture. Most water comes from tributaries within Laos, but dams for hydroelectric power or flood control would reduce the flow downstream.

Cambodia
Hydroelectric power would boost Cambodia's economic development but also displace villagers from fertile land beside the river.

Vietnam
Continuing to dam the Mekong further upstream will reduce the river flow in Vietnam (10% has already been lost).

> **Figure 1** *Dams built and proposed along the Mekong and its tributaries, plus likely impacts on five countries*

Key
- Mekong basin area
- • dam under construction
- ○ commissioned dam
- ● planned dam

Attempts to control globalisation

The theory of globalisation is based on a belief in free flows of people, capital, finance and resources – known as **neo-liberalism**. But neo-liberalism may not always work in practice. Three factors explain why – immigration, censorship, trade.

Limiting immigration

- US President Donald Trump has proposed building a wall along the US–Mexican border.
- Debates in the UK have focused on limiting net migration. This is difficult, with flows from the EU, skills shortages in the knowledge economy and a booming market in overseas university students.

Censorship

In China, the free flow of information and ideas is perceived as a threat to the single-party, communist leadership. The Chinese government enforces the censorship of Internet content, as well as all published material, in order to retain control.

Trade protectionism

In 2016, cheap Chinese steel was being 'dumped' onto global markets. As a consequence, the Indian owners of Tata Steel threatened to close its UK steel plants. A solution would have been to raise tariffs on imported steel (as the USA has done), but this is forbidden by WTO rules.

Maintaining cultural identity

Many cases of resource exploitation in Canada (Figure 2) have caused conflict with traditional communities. The Canadian government has been accused of supporting TNCs against indigenous landholders. In 2013, six out of 21 proposed resource projects were close to collapse, because of protests from traditional communities. Their targets included projects such as:

- oil sands and shale mining (Alberta)
- the Trans Mountain Pipeline between Alberta and Vancouver.

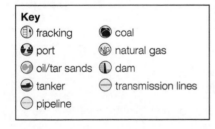

Key
- fracking
- port
- oil/tar sands
- tanker
- pipeline
- coal
- natural gas
- dam
- transmission lines

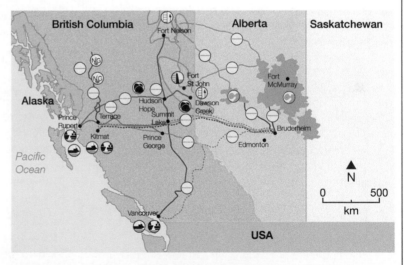

▶ **Figure 2** *The projected pathways of pipelines and electricity transmission lines (called the 'Carbon Corridor') from oil and gas reserves in Alberta and British Columbia to the coast*

 Ten-second summary

- Freedom of movement of people and capital has created culturally mixed societies.
- Tensions can arise as a result of extremism and trans-border water conflicts.
- Censorship, limiting immigration and trade protectionism attempt to control the spread of globalisation.
- Traditional communities often seek to retain their cultural identity.

 Over to you

Create a mnemonic to remember the different ways of controlling globalisation. Make sure you can explain each method in a week's time.

You need to know:

- the costs and benefits of local sourcing
- how Fairtrade and ethical consumption schemes may help the environment and people
- about the role of recycling in managing resource consumption.

Big idea

Ethical and environmental concerns have led to increased localism and awareness of the impacts of a consumer society.

Sustaining globalisation

Many people in HICs get used to having what they want. Global supply chains fulfil these demands and produce is now available in all seasons.

Ecological footprints are one concern (Figure 1). Like most HICs, the UK is living beyond its 'environmental means'. To supply resources for every country at UK levels of consumption would take 3.1 Earths.

▶ **Figure 1** *Who had the biggest ecological footprint in 2013? (data in billion global hectares)*

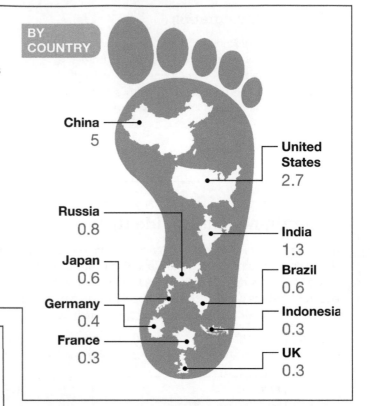

BY COUNTRY

China 5
United States 2.7
Russia 0.8
India 1.3
Japan 0.6
Brazil 0.6
Germany 0.4
Indonesia 0.3
France 0.3
UK 0.3

Sustainable living

Can globalisation and sustainable development exist side by side? The following solutions are intended to make a difference.

Responding locally – Transition towns

Some local groups and non-governmental organisations (NGOs) promote local sourcing of goods to increase sustainability (re-localisation). Totnes in Devon was the world's first '**Transition town**'. '**Transition**' promotes:

- reducing consumption by repairing or reusing items
- reducing waste, pollution and environmental damage
- meeting local needs through local production, where possible.

Advantages of Transition

- Every £10 spent in local businesses is actually worth £23 to the local economy – through the 'Multiplier Effect' (e.g. when local employees and suppliers are paid). In that way, local people gain employment as well as involvement in the local economy.

Disadvantages of Transition

- Strategies like community currencies (e.g. the 'Bristol Pound') threaten global economic growth because they reduce the demand for new items from overseas.
- Most developed countries actually rely on a throwaway culture for their economic growth.
- Some services (e.g. transport) are co-ordinated centrally, so it's hard to influence them.

It has been argued that Transition in a big city like London could be difficult. However, there are currently about 40 community-scale Transition initiatives across London.

Fairtrade

The WTO policy of trade liberalisation can mean that factory workers and commodity growers receive small shares of a product's value (Figure 2). **Fairtrade** aims to return a bigger proportion of the revenue to producers or growers.

Starbucks, for example, claims to help farmers.

- Its Fairtrade Certified Espresso Roast is sourced from small farms in Guatemala, Costa Rica and Peru.
- But, in 2014, only 8.5% of their coffee was Fairtrade certified.

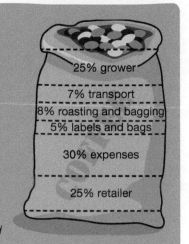

25% grower

7% transport

8% roasting and bagging

5% labels and bags

30% expenses

25% retailer

▶ **Figure 2** *How the revenue from a bag of coffee beans is distributed*

Ethical shopping

The UK's retail sector is increasingly aware of ethical issues in shopping.

- Local produce is returning to supermarket shelves and farmers' markets are commonplace.
- M&S now sells only Fairtrade teas and coffees, plus naturally dyed fabrics in order to reduce carbon emissions.

But there are downsides to **ethical shopping**.

- Buying organic destroys more forests. Less use of fertilisers and pesticides means that more land is needed to produce the same amount.
- Fairtrade increases overproduction, causing prices to fall, which leaves farmers no better off.
- Growing cash crops, even under Fairtrade conditions, can mean that some farmers end up not growing enough food to feed their families.

Waste and recycling

Local authorities manage the disposal of most of the UK's waste. NGOs such as 'Keep Britain Tidy' also try to alter people's behaviour. Since 2004, waste management and recycling in the UK has shown a steady improvement.

- In 2014–15, the total amount of waste recycled was 43.7%, compared to 23% in 2004 (Figure 3).
- In 2000–01, 79% of local authority waste was sent to landfill. By 2013–14, that figure had fallen to 31%.

However, recycling percentages vary.

- In 2014–15, ten councils achieved over 60%.
- But the London borough councils of Newham and Lewisham achieved just 18% in 2013–14.

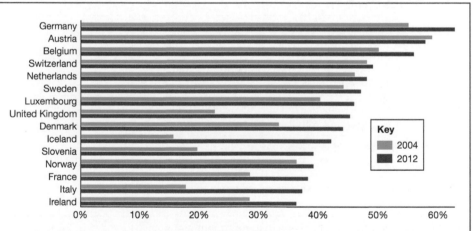

Key
2004
2012

▲ **Figure 3** *Progress in the top 15 European countries for local authority waste recycled and composted, 2004–12*

Chapter 5
Regenerating places

What do you have to know?

This chapter studies ways in which urban and rural regeneration programmes involve a range of players in the processes of both place making (regeneration) and place marketing (rebranding). Regeneration programmes impact on people both in terms of their lived experience of change and their perception and attachment to places.

The specification assumes that you will carry out an in-depth study of the place in which you live or study and one contrasting place; this Revision Guide deals with those contrasting places.

The specification is framed around four enquiry questions:

1 How and why do places vary?
2 Why might regeneration be needed?
3 How is regeneration managed?
4 How successful is regeneration?

The table below should help you.

- Get to know the key ideas. They are important because 20-mark questions will be based on these.
- Copy the table and complete the key words and phrases by looking at Topic 4A in the specification. Section 4A.1 has been done for you.

Key idea	Key words and phrases you need to know
4A.1 Economies can be classified in different ways and vary from place to place.	economic activity (primary, secondary, tertiary and quaternary sectors), types of employment (part-time/full-time, temporary/permanent, employed/self-employed), employment and output data, variations in health, life expectancy and education, inequalities between economic sectors, quality of life indices
4A.2 Places have changed their function and characteristics over time.	
4A.3 Past and present connections have shaped the economic and social characteristics of your chosen places.	
4A.4 Economic and social inequalities change people's perceptions of an area.	
4A.5 There are significant variations in the lived experience of place and engagement with them.	
4A.6 There is a range of ways to evaluate the need for regeneration.	
4A.7 UK government policy decisions play a key role in regeneration.	
4A.8 Local government policies aim to represent areas as being attractive for inward investment.	
4A.9 Rebranding attempts to represent areas as being more attractive by changing public perception of them.	
4A.10 The success of regeneration uses a range of measures: economic, demographic, social and environmental.	
4A.11 Different urban stakeholders have different criteria for judging the success of urban regeneration.	
4A.12 Different rural stakeholders have different criteria for judging the success of rural regeneration.	

You need to know:
- how place identity develops.

Big idea
People become attached to places and identify with them.

Places and big 'events'

Sports events such as the Tour de France or the Olympics often focus on 'special' places (Figure 1). These might include:

- physical characteristics of places, e.g. the Pennines and its steep slopes, challenging Tour de France cyclists
- human characteristics of places, e.g. London's multicultural population as a venue for global events like the Olympic and Paralympic Games.

But places also help to develop a person's sense of identity. There is no such place as 'Yorkshire', but try telling that to someone who feels a strong identity with Yorkshire!

 Figure 1 *Yorkshire pride – the start of the Tour de France in the Yorkshire Pennines*

Developing a sense of identity

People can become strongly attached to places where they live. Essential to becoming attached is a sense of identity. If you 'feel' you're Cornish or Scottish, then you are!

Some factors leading to a sense of identity are personal, e.g. families, friends, in places where we've grown up.

Most factors that make a place distinctive are geographical, e.g.

- physical landscapes, resulting from geology and landscape processes, e.g. beaches, erosion of steep valleys
- human factors such as building style, which is often linked to local rock type (Figure 2)
- economic past, e.g. the industrial factories or housing of many UK cities

Other factors include cultural things such as religion (where places of worship form the heart of places), food and drink (e.g. Yorkshire pudding), and images portrayed in film or TV (e.g. *Coronation Street*, which is about Manchester).

 Figure 2 *Leeds Town Hall – built from local millstone grit and part of the distinctiveness of the city*

Rebranding places

Place identity can be used to encourage economic growth. Deindustrialisation in northern UK cities has caused many to struggle since the 1980s.

- **Regeneration** (redeveloping old industrial areas or housing) can change people's image of a place from being run down to somewhere with economic potential.
- The change of image is called **rebranding** – how places are reinvented and marketed using new identity to attract investors.

Ten-second summary

Place identity arises from factors that are:

- personal (family, friends)
- geographical (e.g. landscape, economy).

Over to you

Draw a spider diagram to explain different factors leading to place identity:

a where you live, and
b in one place you have studied.

You need to know:
- how economic activity is classified
- about changes that have occurred in economic sectors
- how earnings and quality of life vary.

Big idea

Economies are classified in different ways and vary from place to place.

Classifying employment

There are four employment sectors:

- **Primary** – producing crops and raw materials.
- **Secondary** – manufacturing products.
- **Tertiary** – services (e.g. retail, tourism, healthcare, education).
- **Quaternary** – specialist finance, law, IT, biotechnology.

Jobs may also be classified depending on whether they are:

- **full-time** (35 hours per week) or **part-time** (under 35 hours)
- **temporary** or **permanent**
- **employed** or **self-employed**.

Employment change

Increasingly, young graduates are moving to London and south-east England, where opportunities are greater. This has been caused by a change in UK employment sectors since the 1980s:

- a decline in primary and secondary sectors, known as the **old economy**, where jobs halved from 10 million to 5 million between 1980 and 2015
- rapid growth in tertiary and quaternary sectors, known as the **new economy** (or **post-industrial economy**), where jobs increased from 17 million to 28 million between 1980 and 2015.

This results from government policy in the 1980s to change the UK economy from one based on primary and secondary goods to tertiary and quaternary, because:

- Goods produced in the UK were more expensive than those from overseas.
- The global shift of manufacturing to Asia made imported goods cheaper.

As mines and factories closed during the 1980s, well-paid, full-time jobs were cut, especially in northern England, Wales and Scotland. Meanwhile, tertiary and quaternary sectors grew in:

- Tourism and retail (tertiary) – caused by increased leisure and car ownership, and cheaper air travel. These jobs are countrywide but most are seasonal, low paid and part-time
- The **knowledge economy** (quaternary) – high salaried jobs in finance, law and IT for well-qualified graduates. These jobs are **footloose** so can locate anywhere. Banking and finance have located in areas such as London's Docklands (Figure 1), attracted by low tax, good transport and broadband connectivity.

▲ *Figure 1 Canary Wharf in London's Docklands, which was developed to attract quaternary employment in the 1980s and 1990s*

Employment, incomes and inequality

People in different jobs earn very different wages. The mean weekly wage of 'managers, directors and senior officials' (£900) is four times higher than that of workers in 'sales and customer services' (£237). These differences have led to five socio-economic inequalities.

Regional inequalities

Incomes vary regionally (Figure 2). There are more high earners in London because it's the capital, so incomes are higher in quaternary employment and government. Those who work in the Docklands knowledge economy have much higher incomes than average. In London, 58% of jobs are in the three highest income categories (out of nine) and only 22% of jobs are in the lowest three.

Variations in quality of life

Incomes are higher in London and the South East, but housing and other costs are higher. So what about quality of life?

- Personal happiness is based on where people feel happiest, housing affordability and quality of life (Figure 3).
- The happiest places are the most affordable!

Income and life expectancy

Jobs and earnings affect life expectancy. Life expectancy in the highest occupational groups (men 82.5, women 85.2) is 6–8 years longer than in the lowest (men 74, women 78.5).

Income and health

Income affects health. In the 2011 Census, 30% of people on the lowest incomes said their health was 'not good', compared to 13% on the highest.

Variations in educational achievement

The distribution of high GCSE grades varies, as does the percentage of those with university degrees. London has the highest of each (25.3% and 40.5% of people respectively) and north-east England has the lowest (17.6% and 24.3%).

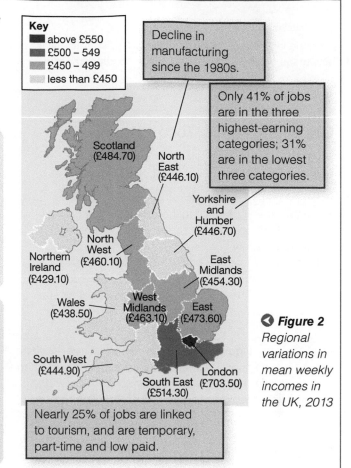

Key
- ■ above £550
- ■ £500 – 549
- ■ £450 – 499
- □ less than £450

Decline in manufacturing since the 1980s.

Only 41% of jobs are in the three highest-earning categories; 31% are in the lowest three categories.

Scotland (£484.70)
North East (£446.10)
Yorkshire and Humber (£446.70)
North West (£460.10)
Northern Ireland (£429.10)
East Midlands (£454.30)
Wales (£438.50)
West Midlands (£463.10)
East (£473.60)
South West (£444.90)
London (£703.50)
South East (£514.30)

Nearly 25% of jobs are linked to tourism, and are temporary, part-time and low paid.

◀ **Figure 2**
Regional variations in mean weekly incomes in the UK, 2013

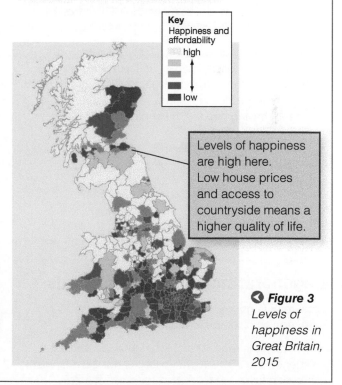

Key
Happiness and affordability
- high
- low

Levels of happiness are high here. Low house prices and access to countryside means a higher quality of life.

◀ **Figure 3**
Levels of happiness in Great Britain, 2015

 Ten-second summary

- UK economy and employment have changed since the 1980s. Primary and secondary have decreased while tertiary and quaternary have increased.
- Different jobs lead to different levels of pay, which creates social inequalities.

 Over to you

1 Explain why London and the South East have gained more from the new economy.
2 Draw a spider diagram or mind map to show how job type affects income, quality of life, health and education.

You need to know:
- how and why East London's functions and characteristics have changed
- how these changes altered the place, its economy and its population.

Big idea
Places can change their functions and characteristics over time.

Global changes, local places

London Gateway lies 30 km downstream from central London. It is the UK's newest port, ready for the world's largest container ships (Figure 1).

- Container shipping led to the closure of the original Port of London in East London because the Thames was too shallow.
- Container ships are needed because of the shift to manufacturing in Asia, which changed the economics of shipping. The biggest ships are cheapest for transporting manufactured goods.

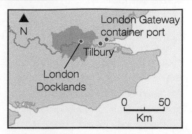

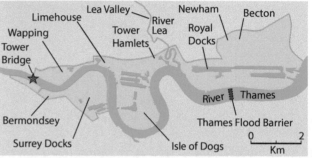

▶ **Figure 1** *Relocation of port facilities from the original London Docklands east of Tower Bridge (in orange on the main map) to the new docks further downstream at London Gateway*

London's changing East End

The last of London's docks (once the UK's largest) closed in 1981. Their closure was devastating.

- Between 1978 and 1983, 12 000 jobs were lost, creating 60% unemployment in Docklands.
- The riverside environment was derelict.
- Nearby, industries in the Lea Valley (Figure 1) closed, having needed the original port for import and export.
- Between 1971 and 1981, the Docklands population fell by 100 000 as people left to seek work.

Re-imaging inner cities

The decline in manufacturing caused similar unemployment in industrial cities such as Leeds.

- Inner-city areas (where industry had been) had a poor image with little potential.
- Falling investment led to declining environmental quality, while crime rose sharply between 1975 and 1985 (burglary up 68%, violent crime up 71%).
- Deprivation and ethnic tensions led to riots in 1981 in many cities e.g. Liverpool (Toxteth), London (Brixton).

The Conservative government reacted by attempting to re-brand inner cities. From 1984, Garden Festivals (and, from 2000, Cities of Culture) were held in old inner cities to improve their image – called **re-imaging**.

Regenerating London Docklands

Closure of London's docks created the potential for regeneration close to Central London.

- Planning the regeneration went to a government agency, the LDDC (London Docklands Development Corporation).
- It aimed to encourage investment from property developers, architects and construction companies.
- The process was called **market-led regeneration** – leaving the private sector to decide the future of Docklands.
- The LDDC focused on economic growth, infrastructure and housing (see Figure 2 and page 107).

⬀ **Figure 2** *New housing in Millwall, used to help re-image London Docklands in the 1980s and early 1990s*

LDDC – regeneration

Economic growth

The flagship was Canary Wharf, now London's second **Central Business District** (CBD).

- High-rise offices (to attract quaternary employment) replaced docks.
- The theory was that wealth from high-income earners would generate other jobs by 'trickling down' to poorer communities.
- Most companies in Canary Wharf work in the knowledge economy, attracting 100 000 commuters daily.
- Employment has grown, but poverty is still present. In 2012, 27% of Newham's working population earned under £7 per hour – the highest percentage of any London borough.

Infrastructure

Accessibility was key to regeneration. New transport **infrastructure** included extending the Jubilee Line and the Docklands Light Railway, and building new roads and London City Airport.

Population and housing

The Docklands population has also been transformed.

- Many older people retired to the Essex coast, replaced by a much younger generation. Newnham's average age was 31 in 2011.
- Large-scale immigration since 2000 has increased the ethnic mix.

Before regeneration, most housing was low cost and rented from local councils. Two changes occurred:

- In the 1980s, the government introduced the Right to Buy, giving council tenants the right to buy their home at reduced price. This reduced the amount of social housing. Half of East London's housing is now privately rented, and demand has raised prices. Lower-income earners have been forced out.
- Dockside warehouses have been **gentrified** and are expensive. Former working-class housing has been bought by the middle classes.

Problems remain

Despite regeneration, high levels of deprivation and poor health persist in Tower Hamlets and Newham (Figure 3).

- The most deprived areas are concentrated in the remaining social housing.
- Housing sold under the Right to Buy programme has not been replaced with affordable housing, so more of people's income is spent on rent.
- Tower Hamlets had London's lowest life expectancy in 2012 (77 years).

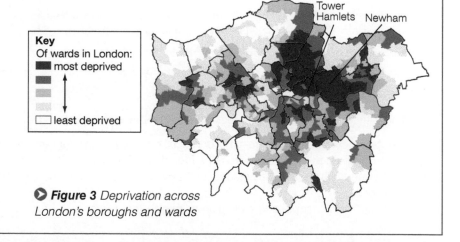

Key
Of wards in London:
■ most deprived

□ least deprived

▶ *Figure 3 Deprivation across London's boroughs and wards*

 Ten-second summary

- The closure of London's docks transformed its functions and communities.
- Government policies regenerated inner cities, transforming East London.
- Population change caused by out- and in-migration changed the character of London Docklands.

Over to you

1 Draw a table to show economic, social and environmental benefits and problems created by the Docklands regeneration.
2 Draw a diagram to assess the importance of the following in changing East London: physical factors, accessibility, government policy, population change.

You need to know:
- how and why places have changed
- how to investigate the characteristics of places.

A sense of belonging

Whether we 'belong' to a place can depend on how many people we know or are related to there. Bethnal Green in East London was studied in the 1950s (Figure 1).

- People there lived close to family members. Often most relatives lived within close proximity.
- The area was industrial and many people worked in manufacturing.
- People shopped locally and worked together in the same factories. Sons worked in the same firms as their fathers, daughters as their mothers.
- Geographers refer to factors binding communities like these as **centripetal forces**.

▶ *Figure 1 Bethnal Green in the 1950s*

Changing communities

Communities like Bethnal Green in the 1950s have disappeared in the UK.

- Changes occurred, often the result of decisions made by people far from the locality.
- Deindustrialisation moved jobs from the factories of East London to South-East Asia.
- Three changes have affected places: globalisation, employment change and inward migration.
- Each has forced people apart – known as **centrifugal forces**.

Globalisation

Globalisation resulted in cheaper imported products from overseas, competing with UK manufacturing and closing London's docks.

- This led to out-migration and the break-up of communities, as people left to find work.
- Populations changed; one population left and another replaced it.
- Bethnal Green's population is now mainly people working in London's knowledge economy, at work all day and without local connections.

Employment change

The change to the new economy has affected East London, where many people are now in high-income professional occupations.

- Many inner urban areas have **re-urbanised**, revitalised by inward migration, gentrification and regeneration.
- But newcomers displaced existing residents, who could no longer afford to live there when house prices and rents increased.
- Regeneration can therefore force people away from places they know.

Inward migration

Economic growth has led to a need for overseas migrants to provide workers. Inward migration changes places, creating new identities.

- Brick Lane in East London (Figure 2) hosted Jewish migrants escaping persecution in the 19th century, and now has a large Bangladeshi population.

▶ *Figure 2* Brick Lane in East London in the 21st century

Investigating change

Investigating changing places is fertile territory for fieldwork.

- Both quantitative and qualitative data help to record changes.
- Quantitative and qualitative data can be **primary** (data you collect first-hand yourself) or **secondary** (data collected by someone else).

Qualitative data

- Qualitative data can include **photos** of places from the present or the past to show changes that have taken place. These can be primary or secondary (e.g. from photo library websites such as Alamy or Getty).
- **Interviews** can record people's experiences about, for example, a sense of community.

Quantitative data

Data can be collected about quality of life (i.e. people's social or environmental wellbeing) using primary data **surveys**.

- Socially, data could be about health, housing or sense of community. You can use a Quality of Life survey either to collect data on different parts of a locality or as a questionnaire to ask others.
- Environmentally, data could be about air quality or noise, for example. Noise can be measured using a decibel meter, available as a free app for smartphones. Environmental quality surveys (EQS) give you a numerical score for several indicators you can use to judge quality of life.

Census **profiling** uses a range of census data to identify population characteristics of an area, e.g. housing, health and education. Local councils publish census profiles about multiple deprivation.

Figure 7 on page 207 of the student book gives the results of an interview used by students in Canning Town.

See Figure 5 on page 206 of the student book for a Quality of Life survey.

See Figure 6 on page 206 of the student book for a Census profile of Canning Town in East London.

Ten-second summary

- Most communities in the UK have changed since the 1950s, as have population characteristics.
- These changes are the result of globalisation, employment change and inward migration.
- The characteristics of places can be measured using quantitative or qualitative data, which may be primary or secondary.

Over to you

Draw two diagrams to show how:

a centripetal forces drew together people in their communities in the 1950s

b how centrifugal forces caused by globalisation, employment change and inward migration have changed places.

You need to know:
- the characteristics of one 'successful' place, and reasons for its success
- why other places face a downward spiral of decline.

Big idea

Some places are more economically successful than others.

Sydney, global city

The population of Sydney is rising fast. It reached 4.8 million in 2015 – an increase of 9% in four years, due mostly to international migration (Figure 1).

- Over 1.2 million British-born people now live in Australia and 1.5 million of Sydney's residents were born overseas.
- It's a multicultural city, with over 250 languages spoken there.

▶ **Figure 1** The Sydney offices of many TNCs have attracted people to live and work in Sydney from all over the world

What makes Sydney 'successful'?

Sydney is part of an economically successful region along Australia's south-east coast, stretching from Brisbane to Melbourne.

- Cities here, like Sydney, have a large proportion of high-income jobs in the 'knowledge economy'.
- In 2015, Sydney was one of the world's 'Alpha' cities. It ranks tenth in the world for quality of life.

Strong economy

Sydney's economy is strong in quaternary employment. Its Gross Regional Product (like GDP, but regionally) was US$337 billion in 2013 – Australia's largest.

- Sydney is the leading financial centre for the Asia-Pacific region.
- It has half of Australia's top 500 companies and two-thirds of Australia's headquarters of global TNCs.
- It has a young workforce (median age 36 compared to the UK's 41).

Attractions for business

- It has an attractive environment (e.g. the Harbour) and climate.
- Its time zone allows business trading in the USA and Europe – essential for banking.
- Government policies encourage globalisation by de-regulating banking (allowing overseas banks to operate there) and inward migration for well-qualified professionals, for example.

High incomes

Australia's average incomes are higher than the UK's.

- Annual adult salaries in 2015 averaged AU$82 000.
- Household salaries – with several earners – averaged AU$145 000 a year.
- However, demand has made property in Sydney extremely expensive.

Living in the sun-belt

Sydney is one of the world's 'sun-belt' cities – a sunny climate has encouraged people to live there. There has been a population drift away from the old industrial cities to sun-belt cities.

- Many choose to live in 'gated communities' – select housing on small estates with security staff.
- Such estates attract high-income earners, paying high service fees.

The spiral of decline

Within the eastern USA are industrial towns and cities such as Clairton and Pittsburgh.

- Many grew because of mining, steel or processing coal for the steel industry.
- The populations of such towns and cities peaked in the 1950s.
- But many now form a region known as the **rust-belt** (Figure 2) because of the decline in metals and coal.
- Their decline has led to **de-industrialisation**.

Reasons for the decline include:

- overseas competition, producing cheaper coal and steel
- mechanisation of coal mines, replacing people
- lower wage costs in south-eastern USA led to the relocation of steel and car industries during the 1990s and 2000s from states such as Michigan to Alabama, Tennessee and Texas.

The US coal industry survives now because of government subsidies without which mines would be forced to close, creating a **negative multiplier**, or downward spiral (Figure 3).

High-income jobs in mining and steel have been replaced with low-wage tertiary jobs in retail and local government.

- Populations of cities have declined, e.g. in Detroit by over 25% from 2000 to 2013.
- Smaller towns in 'coal country', like Beattyville, Kentucky, have many social problems, such as poverty caused by high unemployment.

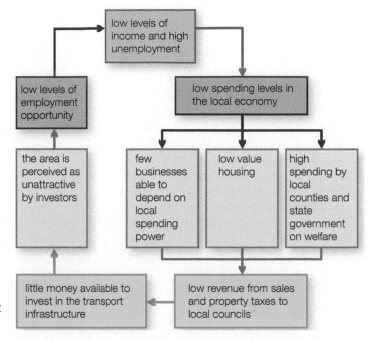

⬥ **Figure 3** *The cycle of decline that threatens the USA's rust-belt*

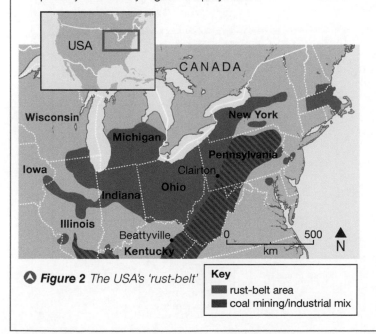

⬥ **Figure 2** *The USA's 'rust-belt'*

Key
■ rust-belt area
■ coal mining/industrial mix

Beattyville – a declining rural settlement

- In 2012, Beattyville's median annual household income was US$12 000 (£8000) – 25% of the national level. Half its families lived below the poverty line.
- In 2013, its population of 1270 lived mainly in trailer homes or log cabins, with a reputation for poverty and crime. 33% of teenagers left high school without graduating and only 5% had college degrees.
- Homelessness and drug crime are rife. In 2013, drug overdoses caused 56% of accidental deaths in Kentucky.
- Men's life expectancy was 68.3 years, eight years below the US average.

 Ten-second summary

- 'Successful' places have a pleasant environment and climate, but also develop because of government policies.
- Places such as the rust-belt have suffered de-industrialisation, with declining job opportunities, falling incomes and social problems.

Over to you

1 Summarise the extent to which Sydney is 'successful', using a SWOT analysis (**S**trengths, **W**eaknesses, **O**pportunities, **T**hreats).
2 Draw a negative multiplier diagram to show economic causes, and social and environmental impacts, of decline in rust-belt cities.

You need to know:
- the ways in which people experience and engage with places
- why this varies.

Big idea
The extent to which people experience and engage with places varies.

Grampound – a sense of a community

Grampound, a village of 800 people in mid-Cornwall (Figure 1), is a strong community. It shows how people can engage with local issues.

- In 2014, its community shop opened, supported by volunteers and financial contributions from the village.
- 25% of the village population is over 65. Social isolation has been a problem for the elderly, who find it difficult to access shops beyond the village.
- A community shop and coffee shop – owned and run by the village – seems to resolve both problems.

Community action

Grampound has a thriving community.

- In 2008, it won a competition to find the UK's best community in the South West, and came second in the UK overall.
- The judges were impressed by its clubs and societies for all ages, and a strong sense of engagement in the village.
- For example, the village raised £50 000 towards the cost of its shop. 90% of households became shareholders, raising £20 500.
- The Prince Charles' Countryside Fund added £19 000. The local Parish Council, rural charities and a renewable energy company added another £10 000.

Figure 1 *Grampound in Cornwall*

The wider significance

Certain factors contribute to Grampound's sense of community and its inhabitants' high level of engagement (Figure 2):

- **Size** – people generally engage less in urban areas.
- **A working village** – people live and work there.
- **Key people** are willing to stand for elections, raise money or organise activities.
- **A range of activities** – despite a small population, Grampound has an annual carnival and 14 clubs and organisations.
- **Politicians** – Grampound's county councillor lives in the village, organises a local produce market and produces monthly newsletters. Residents feel informed.

Grampound people also engage politically.

- Turnout in the 2013 Parish Council Election was 63%, compared to below 30% nationally.
- In 2013, turnout for County Council Elections was 43% (Cornwall's average was 33%).

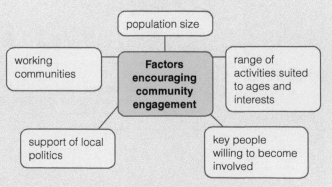

Figure 2 *Factors encouraging community engagement*

Wider engagement

Nationally, more people vote in General Elections than in Local Elections (Figure 3).

- Even then, national turnout has fallen from 82.6% in the 1951 General Election to 66.1% in 2015.
- For EU elections, UK turnout is even lower – only 36% in 2014 compared to 90% turnout in Belgium.
- Political engagement reduces as people feel removed from the centre of power, producing **voter apathy** (Figure 4).

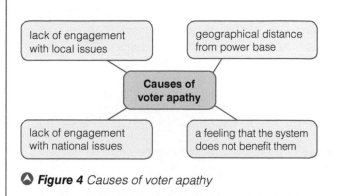

Figure 4 *Causes of voter apathy*

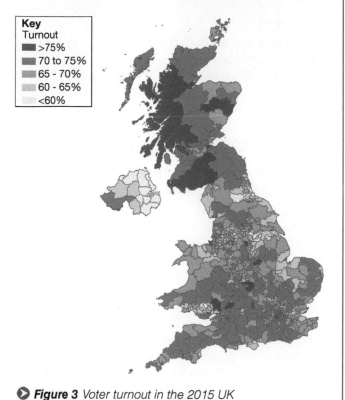

Key
Turnout
■ >75%
■ 70 to 75%
■ 65 - 70%
■ 60 - 65%
■ <60%

▶ **Figure 3** *Voter turnout in the 2015 UK General Election*

How engagement varies

People's engagement varies with:

- **Age** and **gender** – people aged over 60 are more likely to vote in elections; more women engage in community activities than men.
- **Ethnicity** and **length of residence** – places with short-term residents have low levels of engagement. Rural areas have fewer of these than cities, together with fewer ethnic groups.
- **Deprivation** – news website East London Lines claims '*the poor don't vote*'. However, a diverse ethnicity *increases* the likelihood of voting, even in poorest areas. People vote if they have faced prejudice or exploitation, or bring their voting traditions when they migrate.

Regeneration and engagement

Regeneration can reduce engagement, because:

- It is top-down – designed by planners, developers and government, and imposed on people.
- Schemes are measured by economic impacts (measured in sales and land values), not social impacts.
- Groups disagree about who regeneration is for. They may think regeneration only benefits high-income earners.

Regeneration schemes in Cornwall

Support for Cornwall's regeneration schemes varies.

- **Locally**, housing proposals to build new houses in villages are often supported if housing is 'affordable' and incomers help to maintain village services.
- **Regionally**, plans for a waste incinerator in central Cornwall caused protests but Cornwall Council decided to build it. It is regarded as an eyesore and local people fear toxic emissions.
- In line with **national** policies, Cornwall Council supports wind turbines and solar panels even though protestors claim that they spoil the countryside.

 Ten-second summary

- Engagement is more likely to thrive in rural areas and where key people are willing to contribute to communities.
- Political engagement varies according to several factors.

 Over to you

1 Use Figure 3 to identify how UK voter turnout varies geographically.
2 Copy Figures 2 and 4, and add detail to explain how and why engagement varies.

You need to know:

- different data (quantitative) and other evidence (qualitative) that show a need for regeneration
- how the need for regeneration can be seen differently by different players

Big idea

Regeneration takes place according to perceived need.

Regeneration in London

Regeneration is an economic process. But how does it happen?

- Transport is critical. London's Jubilee Line was extended in the 1990s as part of the Docklands regeneration of Canary Wharf.
- Transport improvements provide development opportunities for places en route. They can bring a change of image, e.g. Bermondsey was derelict after London docks closed. The Jubilee Line extension made it a property hotspot as investors bought up old warehouses to gentrify them.
- In the late 2010s, London's Crossrail link created investment opportunities for regeneration to areas along its route (Figure 1).

The process of regeneration

Regeneration involves different **players**, including central government (which funded Crossrail) and private developers (who invest in property re-development).

- The example of Canning Town below shows how different players may prefer a different focus for regeneration.
- Its purpose is often to transform an area socially (e.g. new housing and schools) and environmentally (e.g. creating public spaces). The physical appearance changes as old buildings are demolished and new ones built.

⊘ **Figure 1** *The new Crossrail link to connect East and West London*

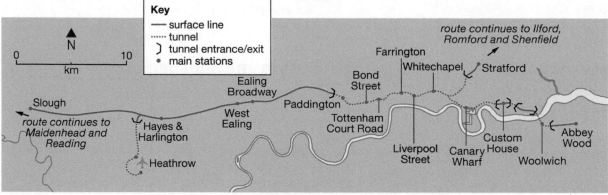

The need for regeneration in Canning Town

Custom House, part of Canning Town in East London, was among London's most-deprived areas in 2001. Figure 2 compares Custom House and Canning Town wards with the rest of London. The needs in Custom House were:

- **Economic** – it needed employment. In 2001, only 37.6% of adults were in full-time work (London average 51.6%) and, in 2011, it still had low incomes and high numbers claiming out-of-work benefits (Figure 2).
- **Social** – it needed improved housing, health facilities and education. In 2001, 71.6% of housing was rented, much of it of low quality. 43.1% of adults had no educational qualifications.
- **Environmental** – docks closure in 1981 created derelict land and there was little public open space such as parkland.

	Percentage of households living in social housing (2011)	Percentage of adults claiming out-of-work benefits (2011)	Median household income estimated (2012/13)
Canning Town North ward	49.3	25.8	28 910
Canning Town South ward	36.0	23.2	32 870
Custom House ward	41.4	22.9	31 840
London as a whole	24.0	12.0 (2010)	39 900

⊙ **Figure 2** *Three indicators of deprivation for Custom House and Canning Town compared to London as a whole*

Deprivation data come from responses to the Census.

- Census data are collated and processed, and then published for wards such as those shown in Figure 2.
- The data are quantitative.
- The data can be used to support qualitative assessments of areas of deprivation, such as local authority reports or people's own descriptions of places.

Regenerating Custom House

The regeneration project is called CATCH.

- From 2001 to 2010, it was funded mostly by central government (Labour), whose focus was **community-led**.
- It aimed to identify community needs and create suburbs with mixed owned and rented housing.

It focused on four aspects:

- **Housing** – 10 000 affordable new homes, particularly family-sized houses planned.
- **Employment** – job and training schemes for local people created; offices and workspaces for small businesses provided; local shops were opened; public transport improved.

- **Education** – buildings for primary and secondary schools improved. By 2015, 59% of Newham's students achieved five or more good grade GCSEs, compared with 27.9% in 1996.
- **Health and community** – a new health centre, library, community centre and children's play areas opened; street safety improved.

However, government (Conservative) funding was replaced by private-sector investment in 2010. So property development now drives regeneration in the Hallsville Quarter in Canning Town.

The Hallsville Quarter

Redevelopment began in 2014, involving property developers, Newham Council and a not-for-profit house-building company (Figure 3).

- It aims to regenerate economically by developing a supermarket, shops, open spaces, bars and restaurants, new homes, small-business premises, a cinema and a hotel.

People's attitudes towards the regeneration vary.

- London's *Evening Standard* newspaper believes that Crossrail will give a huge boost and make property more desirable to high-income groups with greater spending power.
- Already flats cost over £250 000 and four-bedroom houses over £1 million, which creates benefits for property developers.
- Local residents were initially in favour of regeneration, but rents of new properties are high and flats are smaller. Older residents need more help from social services to adapt their homes, and councils have little money.

 Figure 3 *An artist's impression of the new Hallsville Quarter development compared to older housing in the area before regeneration*

Ten-second summary

- The need for regeneration can be measured using census data.
- Needs can be economic, social or environmental.
- Different governments may adopt different styles of regeneration and perceive need in different ways.

 Over to you

1 Explain how each indicator in Figure 2 signals a need for regeneration.
2 Compare the images in Figure 3. Using these and the details provided, outline the benefits and problems of the CATCH regeneration from 2001.

You need to know:

- arguments for and against national regeneration projects like HS2
- how national governments influence regeneration, sometimes conflicting with local communities.

Big idea

National governments are fundamental to regeneration.

HS2 and regeneration

HS2 is the new high-speed rail line that will link London, Birmingham and the North (Figure 1).

- It's controversial. Opponents argue that it will cause environmental damage, while others believe it will upgrade Britain's rail network and regenerate cities like Birmingham.
- The debate is economic need versus environmental cost.

HS2 reverses government transport policies, which have favoured roads rather than railways.

- But the UK has some of Europe's most congested roads, especially in the South East and routes to Birmingham and Manchester (M6), and Leeds (M1). Congestion costs £22 billion annually in lost time.
- Rail travel is also at its highest level, increasing 65% between 2002/03 and 2014. Some routes (e.g. London–Manchester) are close to capacity.

HS2 construction began in 2018, with two phases:

1 a high-speed link (400 km/h) between London and Birmingham, complete by 2025
2 further links to Manchester and Leeds, via the East Midlands and Sheffield, complete by 2030.

▲ **Figure 1** *The projected route for HS2*

Advantages of HS2

Projected benefits include:

- faster journeys between London and Birmingham (cut from 80 to 49 minutes)
- 60 000 construction jobs
- regeneration of cities it links.

Disadvantages of HS2

Opponents believe it will damage Areas of Outstanding Natural Beauty (AONBs) such as the Chilterns and, because there are no intermediate stations between cities, communities along its route will not gain.

Central government funding

HS2 costs £50 billion – too expensive for most private companies to fund without bankruptcy.

- Transport rarely makes a profit, so the government subsidises rail (by £3.8 billion in 2015). Otherwise companies would operate at a loss.

- Central government is funding HS2 because it's seen as an investment, earning franchising fees from train companies to run services and generating economic growth from improved links to the North.
- This will generate jobs and a multiplier effect, from which government receives tax revenue. Therefore benefits outweigh costs.

Government policy – housing

Housing is essential to regeneration so people can live near enough to work. So this big issue also involves government policy.

- With a rapidly rising population, Cornwall, for example, needs 27 000 new homes annually to meet demand, but is only building 16 000.
- Cornwall is a low-wage county, and houses cost 12–16 times average incomes.
- Demand for low-cost social housing is rising fast.
- Should planning restrictions be lifted to allow housing on greenfield landscapes sites or in AONBs?

The causes of the housing demand are:

- rapidly **rising population** (through immigration and high birth rates)
- increase in the **number of households** because of divorce
- **overseas investors** purchasing property as an investment
- demand for **affordable housing** because the Right to Buy scheme (see Section 5.3) reduced social housing stock, leaving shortages.

Government policy – fracking

National government has the power to influence development via local planning laws. This sometimes conflicts with local councils and communities.

- The UK Government supports **fracking** to increase domestic natural gas supplies and improve **energy security**.
- It argues that fracking could encourage regeneration.

But fracking affects rural landscapes such as National Parks (Figure 2) and local opposition to test drilling is fierce. The public are more opposed to fracking than any other energy issue. The reasons include:

- potential subsidence caused by fracturing rock underground
- pollution of acquifers as gas escapes and seeps into rock
- lack of economic benefit, as employment benefits only occur during extraction.

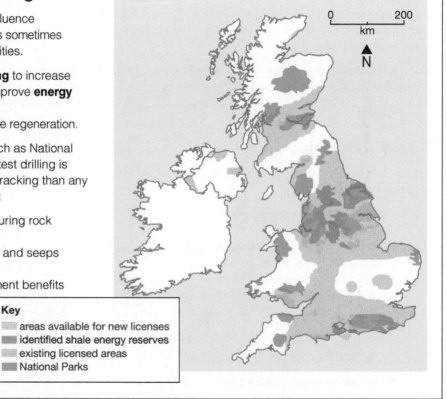

❯ **Figure 2** *Areas in the UK where permission has been granted to search for shale gas by fracking*

Key
- areas available for new licenses
- identified shale energy reserves
- existing licensed areas
- National Parks

Government policy – deregulation

In 1986, the government decided to deregulate the UK's financial sector. This allowed overseas financial companies to set up in London without government approval.

- This led to a huge increase in UK financial services, accounting for 30% of UK GDP in 2015 (double that of 1986).
- It drove the regeneration of London Docklands in Canary Wharf.

Similar government decision-making occurred when the UK joined the European single market in 1992 (allowing free movement of labour).

- This movement helps to balance the UK's ageing population with increased taxation revenue.
- Immigration is controversial for some, while others believe that economic growth requires sufficient labour.

Ten-second summary

- National regeneration projects like HS2 can cause conflict.
- Conflicts are usually economic versus environmental arguments.
- National government becomes involved in local issues, which can conflict with local communities.

Over to you

1 List the costs and benefits of HS2.
2 Explain why central government becomes involved in housing or fracking, which are local issues.

You need to know:

- the economic problems facing rural areas like Cornwall
- how local government tries to attract inward investment.

Big idea

Rural areas can rebrand themselves by attracting inward investment.

Cornwall's economic problems

Cornwall is one of the UK's biggest tourist destinations for both domestic and overseas tourists. That's fine in summer, but it lacks a substantial year-round economy.

- Its 'old economy' consisted of year-round, permanent jobs in the primary sector. Their decline illustrates what geographers call a **post-production countryside** (Figure 1).
- Its 'new economy' varies. The quaternary 'knowledge economy' is small, whilst its biggest industry is tourism, a tertiary activity, which usually involves low-wage, part-time and seasonal jobs.

How might Cornwall develop a new economy with well-paid, year-round jobs to replace its declining primary sector?

Industry	Reasons for its decline
Farming	• Falling incomes, as supermarkets compete for lowest prices for produce like milk • Cheaper imported food • Reduced government subsidies
Fishing	• European fishing boats fish in UK waters • Declining fish stocks due to over-fishing
Tin/copper mining	• Exhaustion of tin and copper reserves • Cheaper overseas competition

Quarrying	• Declining employment in the china clay industry caused by:
	  – overseas competition
	  – use of technology in extraction

⬢ **Figure 1** *Declining employment in Cornwall's primary sector*

Cornwall's isolation

Cornwall is remote from the rest of the UK and not ideal for businesses.

- It lies far from the UK's core economic area (London, the Midlands and northern cities) (Figure 2).
- Journey times by road or rail are long and expensive, and can involve overnight stays as well as travel costs.
- To create jobs, Cornwall needs investment of a type that will help to brand it as a **destination tourism** location.

Re-branding Cornwall

Cornwall's tourism depends on brand. Its assets include:

- **mining heritage** – overseas visitors can trace their ancestors back to tin miners who left in the 19th century
- **heritage** and **literary 'branding'** are often used in TV and film productions (e.g. 'Poldark')
- **scenery**, especially the coast, which has some of the UK's bests surfing beaches
- **food production** of 'Cornish' products, such as pasties, cheeses, ice cream and wines
- **'foodie' restaurants** e.g. started in Padstow by TV chef Rick Stein and at Watergate Bay, Newquay started by Jamie Oliver's 'Fifteen' charity
- **gardens** with sub-tropical plants, which thrive in Cornwall's mild climate
- **outdoor pursuits** such as rock-climbing, surfing and para-surfing, taught at the Extreme Academy at Watergate Bay.

Key
Travel time from Penzance
— by car
— by HGV
— by rail
▨ area in which most UK economic activity occurs
▨ Cornwall

⬢ **Figure 2** *Cornwall's distance from the core area means rural depopulation and fewer job opportunities*

Attracting new investment

Economically disadvantaged areas of the UK qualify for assistance from the UK government and investor incentives from EU funding. To attract investment, Cornwall must compete with other similar areas of the UK for government **Regional Aid** (Figure 3).

Enterprise Zones

Enterprise Zones attract particular types of regional aid. In 2015, there were 44 in the UK, all within the areas on Figure 3. These offer incentives for investors, including:

- council business tax discounts of up to £160 000 a year
- no need for planning permission (except safety building regulations)
- superfast broadband
- tax allowances on the cost of new buildings and training new employees for a few urban Enterprise Zones.

Aerohub Business Park

In 2014, Cornwall Council obtained Enterprise Zone status for Aerohub Business Park near Newquay Airport in a partnership with private investors aiming to diversify Cornwall's economy.

- Its 'brand' is its location, attracting aviation companies.
- It aimed to create 700 permanent skilled jobs by 2015.

By 2015, businesses there included:

- **Aircraft industries** – flight training for pilots, engineers, aircraft maintenance, Coastal Search and Rescue, offshore operations for the RAF and Navy, Cornwall's Air Ambulance, and an aviation museum.
- **Others** – e.g. wind turbine maintenance, development centre for the Bloodhound Super Sonic Car, modular buildings manufacturer.

Impressive, but it generated 450 jobs, few of which were 'new'. Many jobs transferred from elsewhere in Cornwall or were 'displaced' during privatisation of government services.

Enterprise Zones – how successful?

Opinion is divided. Urban locations attract companies, e.g. Warrington (where a science and innovation centre attracted 14 businesses employing 160 people in its first year), Liverpool and Cardiff. Rural areas struggle.

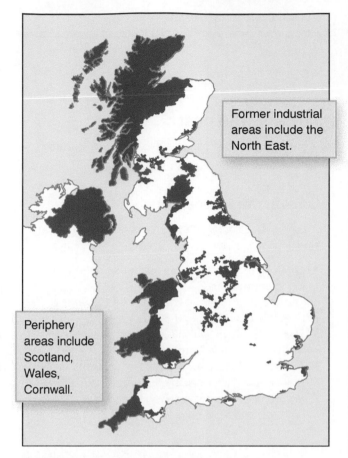

Former industrial areas include the North East.

Periphery areas include Scotland, Wales, Cornwall.

Figure 3 *Areas that qualify for Regional Aid*

 Ten-second summary

- Rural areas such as Cornwall face economic problems caused by the decline of the old economy.
- While the old economy provided permanent, year-round jobs, the new economy provides jobs that are seasonal, temporary and low-paid.
- Attracting inward investment into rural areas is a challenge.
- UK Government investment in rural areas has declined and the benefits of schemes are debated.

 Over to you

1 Draw a mind map to show the problems facing Cornwall's economy and their causes.
2 Draw a table to assess the success of Aerohub Business Park.

You need to know:

- how urban areas rebrand themselves
- how local government policies aim to attract inward investment.

Big idea

Re-branding presents places in attractive ways and attempts to alter public perceptions of them.

Images of the past

Regeneration often uses the past to help 'sell' places. Image is everything.

London Docklands' past

In London Docklands, cranes show the heritage of the area (Figure 1). They're a part of its changing identity.

- London Docklands is now desirable, but it wasn't in the early 1980s.
- The image of the East End was of social unrest (e.g. the 1981 riots) and derelict land.
- The government, which owned the docks, was keen for new development and a new image.
- At the time, local people protested against regeneration proposals by Margaret Thatcher's government. They wanted regeneration to provide better housing and employment for local people.

⬆ **Figure 1** *These cranes in Silvertown no longer serve their original purpose but are left as part of the heritage of London Docklands*

London Docklands' new image

East London's image is now very different. Its economy is growing, it hosted London's 2012 Olympic Games, and it's ethnically diverse. Its new image results from rebranding the past and gentrification.

- Many UK cities have regenerated old industrial buildings at canal- or river-side locations, e.g. Gateshead Quayside.
- Tourism is important, particularly at weekends, e.g. Birmingham's Jewellery Quarter.
- Regeneration has helped to make inner cities desirable places in which to live and work.

Regeneration in Glasgow

Glasgow's past

Glasgow used to be a city based on shipbuilding (Figure 2); now the River Clyde has just three remaining shipyards. Shipbuilding supported an industrial region based on related industries (engineering, steel, coal). When cheaper overseas competition led to the collapse of Glasgow's shipbuilding, other industries fell with it, known as 'the **domino effect**'. The collapse led to a downward multiplier, with falling revenue for shops and services.

⬆ **Figure 2** *The Titan Crane on Clydeside was once used in shipbuilding but is now a visitor attraction*

Glasgow's new image

Glasgow's regeneration has been funded by the City Council and the Scottish government. Since 2000, tertiary and quaternary industries have grown in arts, culture, sport and tourism.

- The aim is to establish Glasgow as a tourist destination. It was European Capital of Culture in 1990, UK City of Architecture and Design in 1999, and hosted the 2014 Commonwealth Games.
- Its is world famous for its architecture and the Burrell Collection (art museum). The Scottish Exhibition and

Conference Centre, museums, and the Titan Crane (a refurbished shipyard crane) all attract visitors (Figure 2).

- Tourism brings job opportunities in hotels, bars, restaurants and retail.

Further regeneration includes:

- residential development along the Clyde, including shops and restaurants.
- media industries, e.g. BBC Headquarters for Scotland and commercial broadcaster STV.

Regeneration in Plymouth

Kick-starting change

In the 1960s, the naval city Plymouth was re-developed after extensive bombing in the Second World War. Its pedestrianised streets were innovative. Unfortunately, its economy has declined and the city centre now looks dated. Its naval shipyards have been reduced and it competes with Portsmouth to keep ship repair and servicing. Plymouth's remoteness makes investment hard to attract.

 Figure 3 *Royal William Yard, Plymouth*

Regeneration projects

Plymouth City Council plays a key role in regeneration projects such as:

- Drake Circus shopping centre in the CBD, designed to attract people and create jobs. Some feel it takes business away from other parts of Plymouth
- cruise liner terminal. Plymouth's history encourages tourism (e.g. its links to the Pilgrim Fathers in 1620), but only 26 cruise ships visited in 2014
- Science Park, linked to Plymouth's universities and teaching hospital, and designed to develop the knowledge economy. 70 businesses employ 800 people in marine engineering, medicine and renewable energies
- Royal William Yard (Figure 3), a Grade 1 listed former Navy supply store from the Napoleonic Wars. Restoration has taken 20 years and includes shops, restaurants and over 200 apartments. It's near the new cruise liner terminal, but is 2 km from the CBD, a disadvantage for businesses.

Ten-second summary

- Urban areas such as Glasgow have suffered from de-industrialisation.
- Urban areas rebrand themselves using retail, commercial, residential and environmental improvements.
- Rebranding is a strategy designed to create inward investment.
- Local governments try to attract inward investment from private investors.

 Over to you

Draw a Venn diagram with three circles (Docklands, Glasgow and Plymouth).

a On it, identify features of each regeneration programme, highlighting economic, social and environmental achievements.

b Assess the most and least successful outcomes of each regeneration programme.

You need to know:
- why Barking and Dagenham need regeneration, and where
- the criteria by which regeneration success can be judged.

Big idea
The success of regeneration depends on the criteria used to judge it.

Changes in Barking and Dagenham

The East London borough of Barking and Dagenham once had one of Europe's largest car factories, employing 40 000 people. However, Ford closed this factory in 2002. It still produces diesel engines, but most work is automated so Ford now employs 3200. In 2013, Sanofi (a pharmaceutical TNC) closed its operation there too.

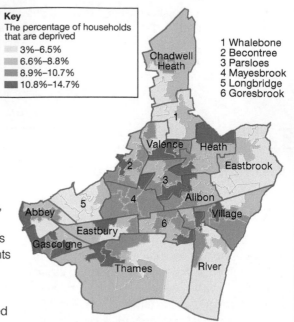

Key
The percentage of households that are deprived
- 3%–6.5%
- 6.6%–8.8%
- 8.9%–10.7%
- 10.8%–14.7%

1 Whalebone
2 Becontree
3 Parsloes
4 Mayesbrook
5 Longbridge
6 Goresbrook

Deprivation in Barking and Dagenham

The closures have increased deprivation: economic (unemployment), social (poor health) and environmental (derelict land). By 2015, the borough was the ninth most deprived part of England. It had London's highest adult unemployment rate (9.8% of adults) and 27% of residents earned less than London's living wage.

Figure 1 shows deprivation levels in small areas called **Lower Super Output Areas (LSOA)**. The darkest shading shows deprivation caused by at least three causes:

- **unemployment** or long-term sickness
- **low qualifications** at GCSE
- **health and disability**, i.e. long-term health problems
- **poor housing**, e.g. overcrowded, shared or without central heating.

⬆ **Figure 1** *Deprivation levels in Barking and Dagenham: the named areas are council wards (local districts)*

Measuring deprivation

Indicators of deprivation can be:

- **Economic** – e.g. low incomes. In 2013, 29% of jobs in Barking and Dagenham were high-income jobs compared to 50% in London as a whole.
- **Social** – including low educational attainment and skills, health and disability, crime risk and unaffordable housing.
- **Environmental** – poor environmental quality (e.g. air quality).

See page 233 of the student book for a table showing seven deprivation indicators for different parts of London.

Regeneration in Barking and Dagenham

Regeneration should reduce deprivation. There are several suitable sites for regeneration in Barking and Dagenham, all with excellent tube and rail links to Central London. They include:

- **Gascoigne Estate** – is the most deprived area (see Figure 2). Plans include 1550 houses, schools, a community centre, work and leisure spaces by 2024.
- **Dagenham Dock** – an old industrial site with derelict land, and fuel and chemical tanks. It is now a sustainable business area with a recycling centre and an anaerobic digestion plant.

▷ **Figure 2** *Five regeneration sites in Barking and Dagenham*

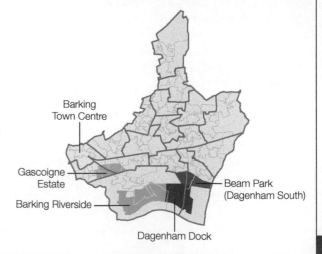

Barking Town Centre

Gascoigne Estate

Barking Riverside

Dagenham Dock

Beam Park (Dagenham South)

- **Barking Riverside** – the site of a former power station. Plans include 11 000 houses, schools, work and leisure spaces, and 6000 jobs by 2020.
- **Barking Town Centre** – was run-down. By 2014, there were 400 new houses and 1000 m² of new commercial space.
- **Beam Park (Dagenham South)** – this old Ford site creates space for 40 000 m² of workplaces, and a hotel and pub already employ local residents.

For more detail about these regeneration sites, see page 234 of the student book.

Measuring the success of regeneration

The success of regeneration can be judged using four criteria:

- **Economic** – e.g. increasing average incomes and the number of jobs (especially better paid).
- **Social** – e.g. reducing levels of deprivation.
- **Demographic** changes – e.g. improving life expectancy and reduced health deprivation.
- **Environmental** – e.g. reducing pollution and the amount of derelict land.

Figure 3 shows how difficult it is to measure impacts of regeneration. It compares data in 2013 for Barking and Dagenham with two other boroughs where regeneration has occurred (e.g. Tower Hamlets contains Canary Wharf).

- Annual % change is the change in number of jobs from 2012 to 2013. Barking and Dagenham had the largest increase even though deprivation there is worse.
- Income percentiles are the lowest earning 10%, 25%, etc. and the 80 percentile is the highest 20%.
- Notice how Newham has lowest incomes in spite of regeneration in the Olympic Park. Incomes in Newham actually fell!

It is equally difficult to measure the impacts of regeneration on health. It takes 20–30 years to see whether changes occur, and even then it is hard to know whether these are linked to regeneration.

Name	No. of jobs (000)	Median income, 2013 (£)	Annual % change	Income percentiles				
				10 (£)	25 (£)	40 (£)	60 (£)	80 (£)
Barking and Dagenham	41	517.40	6.0	150.40	304.90	446.10	604.60	806.60
Tower Hamlets (contains Canary Wharf)	217	804.90	5.0	333.50	535.50	690.40	951.10	1449.60
Newham (contains Olympic Park)	63	475.70	-5.8	122.40	277.70	395.60	550.70	736.20
UK	24 385	403.90	0.0	118.60	244.40	339.90	479.30	689.90

 Figure 3 *Employment and income data in Barking and Dagenham in 2013*

For a comparison of health indicators for Barking and Dagenham, see Figure 5c on page 235 of the student book.

 Ten-second summary

- Regeneration is needed in Barking and Dagenham.
- There are several sites where regeneration might take place.
- It is difficult to measure the impacts of regeneration.

 Over to you

1 For Barking and Dagenham, write down:
   a three reasons why regeneration is necessary
   b three reasons why the success of regeneration can be hard to determine.
2 Study the details about Barking's five locations for regeneration. Assess which you think are the best two sites for regeneration.

You need to know:
- who the players are in urban regeneration
- how different players have different criteria for judging the success of urban regeneration.

Big idea

Different players have different criteria for judging the success of urban regeneration.

A scene of change

Regeneration has long been part of East London, from the Docklands regeneration of the 1980s and 1990s to London's 2012 Olympics, and the construction of Crossrail more recently.

- Projects vary in terms of costs, benefits, who carries them out, who pays, and their success.
- Investment may come from the private or public sectors, or partnerships between the two.

▶ **Figure 1** *Stratford's Westfield shopping centre in East London*

Private-sector investment

This includes Stratford's Westfield Centre (Figure 1), Europe's largest shopping centre and an example of **retail-led regeneration**.

- It created over 10 000 new jobs.
- Westfield, which owns 50%, borrowed £700 million to build it. It recovers this by leasing space to shops.
- In its first four years, the Centre's annual turnover was £1 billion, adding to the local economy.

Public-sector investment

This includes London's 2012 Games.

- The UK government bid for the Games was supported by London's Assembly and its Mayor.
- The Games cost £9.3 billion, recovered through ticketing, TV sponsorship and the sale of property afterwards. It made a profit.

Public-private partnerships

This includes London Docklands regeneration.

- The government handed over land and grants, while property developers created jobs and built offices and housing.
- Although costs were high, the government saves on reduced unemployment and earns tax revenue from jobs.

Which is best?

The pros and cons of private and public sector regeneration vary according to each sector's aims.

Partnerships occur where investment is expensive (e.g. Crossrail) and the government helps to bear the cost.

- The private sector values economic regeneration, when investment is affordable and will return a profit.
- The public sector sees public benefit as important, e.g. affordable housing and open spaces for people.

For an analysis of private and public sector regeneration, see Figure 3 on page 237 of the student book.

London's 2012 Olympic Games

Despite previous regeneration, East London in the early 2000s contained some of the UK's poorest areas. The 2012 Games were intended to bring **convergence** (reduce the gap) between London's poorest and wealthiest areas. They were designed to:

- improve East London's 'brand' as a place to visit, live and work
- attract investment.

Legacies from the Games include:

- the Queen Elizabeth Olympic Park – 560 acres of parkland and employment (Figure 2)
- the Olympic venues, e.g. the London stadium (now West Ham FC), Aquatics Centre (now a local swimming baths), Velodrome (still hosting cycling events) and Here East, the Media Centre (now a base for technology companies)
- the 'International Quarter' (now offices and a 4-star hotel)
- residential areas, providing 9000 new homes by 2025.

 Figure 2 *Regeneration in East London – Stratford's Westfield Centre (foreground) and the Queen Elizabeth Olympic Park and London Stadium (background)*

Further detail about the 2012 Games, the players involved and their impact can be found on pages 237–239 of the student book.

Key players in the 2012 Games

- **UK central government** – the London Legacy Development Corporation is an appointed (unelected) central government agency overseeing the legacy development of the Olympic Park. Its aim is to generate employment and (mostly private) housing.
- **Regional government** – e.g. the London Assembly, which wants to see (mainly affordable) housing and economic growth in East London.
- **Local government** – Four elected London borough councils hosted the Games, and all want economic and social regeneration to continue post-2012.
- **Environmental stakeholders** – charities such as the London Wildlife Trust has cleaned up and re-landscaped what was derelict land, and now includes wetlands with rising numbers of species such as newts, fish, bats and birds.
- **Stakeholders in local people** – especially in housing. London property is extremely expensive, and affordable housing is needed for those on low incomes.

 Ten-second summary

- Players involved in regeneration may be public or private sector, or partnerships between the two.
- The decision about whether regeneration schemes involve the private or public sector, or partnerships, depends on the purpose and cost of each project.
- Different players have different criteria (economic, social, environmental) for judging the success of regeneration.

 Over to you

1 Summarise differences between public- and private-sector investment.
2 Explain why the public, not private, sector funded London's 2012 Olympic Games.
3 Explain why different players might judge the economic, social and environmental impacts of the 2012 Olympic Games differently.

You need to know:

- who the players are in rural regeneration
- how different players have different criteria for judging the success of rural regeneration.

Big idea

Different players have different criteria for judging the success of rural regeneration.

Changing times

Despite its geographical disadvantages (see Section 5.9), Cornwall's economy grew faster than the UK average between 2000 and 2010. Investment increased jobs and boosted tourism.

- However, most investment came from the UK government and the EU.
- Since 2010, cuts in government spending mean that investment is needed from the private sector.
- Private sector investment is hard to achieve where rural population density is low.

Rural disadvantage

The rural economy faces challenges.

- In 2011, Cornwall had England's lowest average earnings, 77% of the UK average.
- 20% of Cornwall's working population earned less than £7.45 per hour.
- Lack of investment means a lack of opportunity in Cornwall.
- Cornwall's young, well-qualified residents leave to find work, causing a 'brain drain'.

Funding regeneration

Before 2010, public sector investment regenerated Cornwall. Until 2007, EU funding called Objective One was designed to raise rural incomes. It worked by **match funding** (matching capital raised by individuals) to **pump-prime** business.

One farm shop proposal, whose owners invested £20 000, worked as follows:

- Their bank lent them £20 000 (so, £40 000 total).
- Cornwall Council matched the new amount (making £80 000).
- The South West Regional Development Agency matched that (£160 000).
- EU Objective One matched that, creating £320 000.

Within 10 years, the farm shop's annual turnover was £700 000, employing 20 people. By 2007, Objective One had invested £230 million in 580 projects in Cornwall.

Key players in Cornwall's regeneration

- **The EU** through Objective One.
- **UK central government** invested through Regional Development Agencies until 2010. These have now been abolished.
- **Cornwall Council** has had no start-up funding since 2010. It offers rebates on business taxes in its Enterprise Zone (see Section 5.9).
- **Environmental stakeholders**, which promote Cornwall's environment (e.g. National Trust) or its potential for renewable energies.
- **Stakeholders in education**, e.g. Combined Universities in Cornwall.

Cornwall's regeneration hotspots

Superfast Broadband

By 2016, over 95% of Cornwall had fibre broadband, the world's largest rural fibre network. It was granted £53.5 million from the EU and £78.5 million from BT, benefiting tourism, knowledge-economy companies and home workers.

For more detail about regeneration projects in Cornwall, and the players involved, see pages 242–243 of the student book.

Figure 1 *Advertising used for Superfast Broadband in Cornwall*

Watergate Bay, Newquay

The Extreme Sports Academy offers courses in surfing and its owners run the local hotel, employing 50–60 people. Next door is the restaurant Fifteen, which trains local young people from disadvantaged backgrounds in catering.

Combined Universities in Cornwall (CUC)

CUC was created in Falmouth to offer degree courses, help graduates set up businesses and create jobs in Cornwall's small knowledge economy, reducing Cornwall's 'brain drain'. Falmouth's 'student economy' has created a healthy property rental market and an evening economy of bars and restaurants.

Wave Hub

A trial wave-power project 16 km off Cornwall's coast, the hub is installed on the seabed, acting as a 'socket' for four different wave-energy converters to plug into and take electricity to the mainland. Its capacity is 20 MW (the same as 6–8 wind turbines). It cost £42 million. If it works fully, it will earn £76 million over 25 years and create 170 jobs.

The Eden Project

A tourist attraction, this now consists of two large 'biomes', an education centre and a hostel for residential groups (Figure 2). It was developed from an old china-clay quarry (Figure 3). It cost £140 million; 75% from the Lottery, the government and the EU. It has generated £2 billion for the Cornish economy, attracted 15 million visitors and employed 650 people.

 Figure 2 *The old china clay quarry in which the Eden Project was built*

 Figure 3 *The Eden Project*

More detail on the Eden Project can be found on page 243 of the student book.

 Ten-second summary

- Regeneration has suffered from government cuts and withdrawal of EU funding.
- Rural areas find it hard to attract investment compared to urban areas.
- Different criteria can be used to judge the success of regeneration. These can be economic, social or environmental.

Over to you

Assess each of the five regeneration projects, based on:

- how well it suits Cornwall's environment
- its economic, social, environmental costs and benefits
- its effect on employment
- whether it develops a 'knowledge economy' and helps to prevent 'brain drain'.

Chapter 6
Diverse places

What do you have to know?

This chapter studies ways in which local places vary both demographically and culturally with change driven by local, national and global processes. These processes include movements of people, capital, information and resources, making some places more demographically and culturally varied while other places appear to be less dynamic.

The specification assumes that you will carry out an in-depth study of the place in which you live or study and one contrasting place; this Revision Guide deals with those contrasting places.

The specification is framed around four enquiry questions:

1   How do population structures vary?
2   How do different people view diverse living spaces?
3   Why are there demographic and cultural tensions in diverse places?
4   How successfully are cultural and demographic issues managed?

The table below should help you.

- Get to know the key ideas. They are important because 20-mark questions will be based on these.
- Complete the key words and phrases by looking at Topic 4B in the specification. Section 4B.1 has been done for you.

Key idea	Key words and phrases you need to know
4B.1 Population structure varies from place to place and over time.	uneven population growth in the UK, population structure, population density, the rural-urban continuum, accessibility, differences in fertility and mortality rates, international and internal migration
4B.2 Population characteristics vary from place to place and over time	
4B.3 How past and present connections have shaped the demographic and cultural characteristics of your chosen places.	
4B.4 Urban places are seen differently by different groups because of their lived experience of places and their perception of those places.	
4B.5 Rural places are seen differently by different groups because of their lived experience of places and their perception of those places.	
4B.6 There is a range of ways to evaluate how people view their living spaces.	
4B.7 Culture and society is now more diverse in the UK.	
4B.8 Levels of segregation reflect cultural, economic and social variation and change over time.	
4B.9 Changes to diverse places can lead to tension and conflict.	
4B.10 The management of cultural and demographic issues can be measured using a range of techniques.	
4B.11 Different urban stakeholders have different criteria for assessing the success of managing change in diverse urban communities.	
4B.12 Different rural stakeholders have different criteria for assessing the success of managing change in diverse rural communities.	

You need to know:
- how and why diversity varies in different parts of the UK.

Big idea
Diversity varies in different parts of the UK.

Big city, big changes

Every year, students leave home for universities in London and other large cities. Many notice differences between the lifestyle in a rural area or small town and that in a densely populated city. The population of a single London borough (there are 32) can be as high as a city like Plymouth, and city population densities of 14 000 people per km² can be 100 times greater than in rural areas.

Rural vs urban populations

The differences between rural and urban populations are stark (Figure 1).

- 49% of Tower Hamlets' residents are aged 20–39. Its median age (29) is Britain's lowest. London's median age is 33, the UK's is 39.4 and East Devon's is 47.
- Many people retire to East Devon. 30% of its population is over 65, compared to Tower Hamlets 6%, London 11% and England 16.3%.

Equally stark is the difference in **ethnic mix**.

- Over 93% of residents in East Devon classed themselves as White British in the 2011 Census, compared to 45% in London and 31.2% in Tower Hamlets. Figure 2 shows other contrasts.
- Ethnic differences lead to greater **diversity** of shops and services within communities.

Cities such as London offer faster economic growth, jobs for graduates and higher salaries compared to other parts of the UK. London's **knowledge economy** (financial services, law, the media) absorbs many graduates. It has a younger, more diverse workforce. Job vacancies exceed numbers of skilled people in the UK, so employers also recruit overseas.

Two processes affect Tower Hamlets:

- internal migration of young graduates from other parts of the UK
- international migration from overseas.

East Devon is also affected by two processes:

- internal migration of retired people from other parts of the UK
- out-migration of younger people to cities such as London.

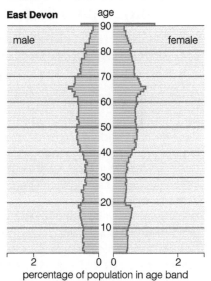
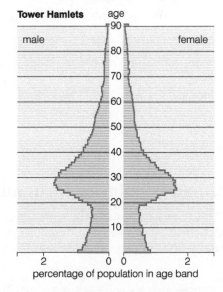

Figure 1 Age-sex diagrams for East Devon and Tower Hamlets

Place	% White British	% Mixed	% Asian or Asian British	% Black or Black British	% Chinese
England and Wales	83.4	1.80	5.87	2.81	0.82
Tower Hamlets	31.2	2.85	30.58	6.47	1.58
East Devon	93.2	0.90	1.51	0.68	0.30

Figure 2 Ethnic diversity of East Devon, Tower Hamlets, and England and Wales

Ten-second summary

- Diversity varies between rural and urban areas of the UK, with younger, more diverse, well-qualified people in cities.

Over to you

1 Write down four differences between the populations of East Devon and Tower Hamlets.
2 Explain why diversity varies in different parts of the UK.

You need to know:

- that population growth in the UK varies regionally
- how and why population structure varies from place to place and over time.

Big idea

Population structure varies from place to place and over time.

UK population density

The population density of the UK varies widely.

- The UK has an overall density of 267 people per km².
- London has 5500 people per km².
- England is more densely populated (413 people per km²) than Wales (149), Northern Ireland (135) or Scotland (68).

- Urban areas account for 89% of UK population, but only 7% of its area. 93% of the UK is therefore rural to some degree.

A growing population

The UK's population is increasing rapidly, especially in major cities like London. This presents challenges. Where will everyone live? How can services be provided?

Population growth varies regionally (Figure 1) because of the following factors:

- **Economic** – London's expanding knowledge economy has led to inward migration of qualified workers and families. Compare the North East, where collapse of traditional industries (coal, steel, engineering, shipbuilding) has led to outward migration.
- **Demographic** – **Internal migration** has attracted young graduates to London, along with **international migration**.
- **Social** – longer life expectancy, caused by falling **mortality rates** amongst over 65s due to increased cancer survival and care of the elderly. In London, male life expectancy rose seven years from 1993 to 2013.

Key
Percentage growth rate
- ■ 20 or above
- ■ 15–20
- ■ 10–15
- ■ 5–10
- ■ 0–5
- ■ below 0

▶ **Figure 1** *Population growth rate projections for the nine English regions, 2012–2022*

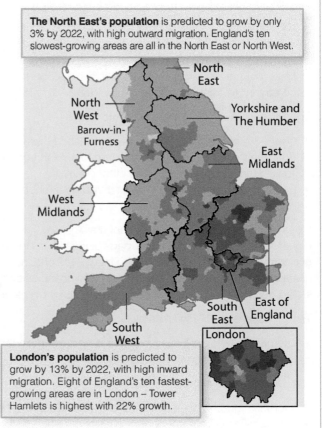

The North East's population is predicted to grow by only 3% by 2022, with high outward migration. England's ten slowest-growing areas are all in the North East or North West.

North East
North West
Yorkshire and The Humber
Barrow-in-Furness
East Midlands
West Midlands
South East
East of England
South West
London

London's population is predicted to grow by 13% by 2022, with high inward migration. Eight of England's ten fastest-growing areas are in London – Tower Hamlets is highest with 22% growth.

Urban population structure

The UK population varies by density and structure (i.e. age-sex). This can be illustrated by Newham and Kingston-upon-Thames (Figure 2), rural-urban fringe, North Yorkshire and the Scottish highlands.

Newham, inner East London

- Dominated by 21–40 year olds. Average age 31.
- Rapidly growing population due to **natural increase** and internal migration of graduates.
- High international migration since 2000. By 2011, 55% were born overseas.

Kingston-upon-Thames, outer South-West London

- Dominated by families. Average age 37.
- Most are UK-born; 20% were born overseas.
- One of London's wealthiest boroughs, as high-income couples move there to raise families.

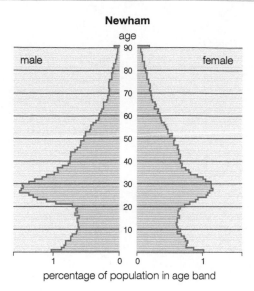

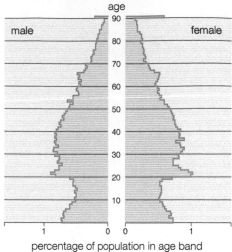

 Figure 2 *Age-sex structure of London boroughs Newham and Kingston-upon-Thames*

The rural–urban continuum

Britain's population becomes more rural with increased distance from cities.

For a model showing the **rural-urban continuum** to represent a rural-to-urban spectrum, see Figure 4 on page 251 of the student book.

The rural-urban fringe

- The **rural-urban fringe** is a blurred boundary between cities and rural areas found at the edge of most cities.
- As urban areas expand, they absorb open countryside by **urban sprawl**.
- Transport routes turn accessible settlements into **suburbanised** places that adopt shops and services.
- Further from the city, these give way to more remote villages and sparsely populated rural areas.

Mixed rural areas, e.g. North Yorkshire

- These areas are mainly rural, but vary between remote uplands (the Pennines) and urban areas (e.g. York, Scarborough).
- The population is ageing (average age 39.8), due to falling mortality rates.
- Remote upland areas are losing younger people. The challenge is to encourage new jobs and housing.
- Populations of towns are increasing, due those seeking **accessible** country living and retirement.
- International migration is small; less than 5% were born overseas.

Remote rural areas, e.g. Scottish Highlands

- These are sparsely populated with remote hamlets and farms far from neighbours.
- The population is ageing (average age 43.2) but increasing rapidly due to internal migration by retired people and families seeking a rural lifestyle.
- There is little international migration.
- Accessibility is a problem. Mountains, lochs and severe winter weather prevent daily commuting to Scottish cities.
- The economy is mainly tourism and farming.

 Ten-second summary

- The population of the UK is rising because of economic, demographic and social reasons.
- Urban areas tend to be younger, attracting young graduates and migrants.
- Rural areas vary in nature, from rural–urban fringes to remote areas.

 Over to you

Draw up a table to compare population characteristics of the five areas (Newham, Kingston, rural–urban fringe, North Yorkshire and Scottish Highlands) discussed in this section.

You need to know:

- why population characteristics vary from place to place
- how different levels of cultural diversity can be explained
- how migration changes population characteristics.

Big idea

Population characteristics vary from place to place and over time.

Urban communities

Communities often emerge in cities just as much as in rural areas. Local suburban coffee shops and pubs are part of the communities that thrive in cities.

- Southall in West London shows evidence of a long-established Indian community.
- Large-scale Indian immigration to the UK began after India's independence in 1947.
- Migrants settled in Southall, many of them Sikh soldiers who were in the British Army. Many were middle-income farming families from Punjab state.
- Some gained employment in London's factories and service industries. Others found jobs at Heathrow Airport.
- Now, half of Southall's residents are Asian or Asian British (Figure 1).
- Other Indian communities settled elsewhere in London. Asian or Asian British residents with Pakistani heritage live in different parts of London.
- Concentrations of communities like these are called **ethnic enclaves**, together creating London's **cultural diversity**.

Urban cultural diversity

Most large UK cities have **social clusters** of ethnic communities.

- The Hindu community in Wembley began to grow when local specialist food shops, places of worship, community centres and restaurants gave support against racism in the 1960s and created a shared identity.
- Now, clusters like this have created economic clusters such as Bradford's 'curry mile'.

A

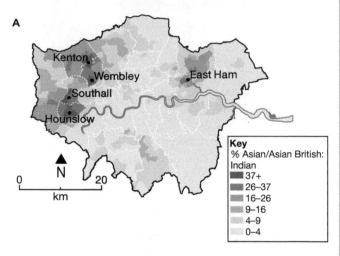

Key
% Asian/Asian British: Indian
- 37+
- 26–37
- 16–26
- 9–16
- 4–9
- 0–4

B

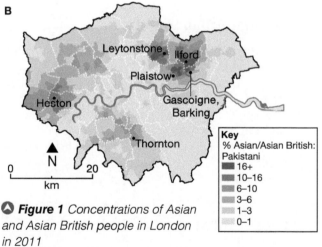

Key
% Asian/Asian British: Pakistani
- 16+
- 10–16
- 6–10
- 3–6
- 1–3
- 0–1

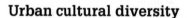
Figure 1 *Concentrations of Asian and Asian British people in London in 2011*

Understanding UK immigration

The UK has a long history of immigration.

- South Asian migrants came after 1947 to escape war, seek opportunity or join family members.
- Where they lived was determined by accessibility to key cities where there was employment. Many Pakistani migrants gained employment in northern textile cities (e.g. Bradford) or in car manufacturing cities (e.g. Birmingham, Luton).
- Remote rural Britain offered few opportunities, and remains mostly White British today.

Many came when the government advertised overseas to fill job vacancies in factories, transport and the NHS.

- Afro-Caribbean migrants arrived from 1948, settling in areas such as Brixton.
- London's annual Notting Hill Carnival marks the contribution made by its West Indian community.

Government policy and EU migration

Since 1995, numbers of overseas-born people in the UK have doubled.

- The main reason is EU membership since 1995 when free movement of EU citizens was introduced.
- With the EU's second largest economy, UK migration was always likely to be high.
- Major cities gained greatest numbers. London's dominance of the UK economy means that half of all overseas-born UK residents are in the South East region.

The top ten source countries for UK migrants include four EU members (Poland is the largest).

- Most migrants are aged 21–35, with equal numbers of skilled and unskilled workers.
- **Skilled migrants** work in London's knowledge economy. Skills shortages force companies to recruit overseas. Most tend to be qualified professionals from the EU, USA and Australia.
- **Semi- and unskilled migrant workers** fill job shortages in refuse collection, childcare, construction, hotels and restaurants. Many are from the EU, but also southern Asia and West Africa.
- Some EU migrants live in rural areas, working on farms.

Impacts of immigration

The UK's population is ageing.

- Between 1974 and 2014, the UK's median age increased from 33.9 to 40.0.
- In 1974, 14% of the population was aged 65 or more. By 2014, it was 18%.
- Changes are caused by increased longevity, decreasing birth rates, falling total fertility rates (Figure 2) and falling mortality rates (Figure 3).
- Increased prosperity leads to falling birth and fertility rates as women make career choices, marry later and have fewer children.
- Falling mortality rates result from improved treatment of diseases such as cancer.
- This leads to a smaller percentage of workers, which increases the **dependency ratio**. Immigration rebalances this because migrants tend to work, increasing government tax revenue to support the non-working population.

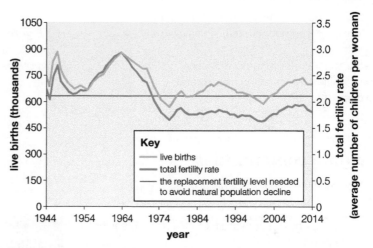

 Figure 2 Changes in the UK's live birth and total fertility rates, 1944–2014

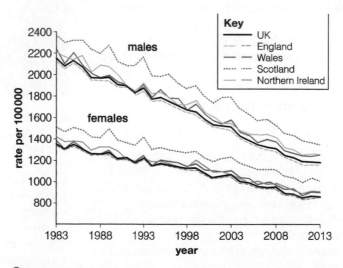

Figure 3 Changes in the UK's mortality rates, 1983–2013

Ten-second summary

- UK urban populations have increased and changed over time.
- Ethnic enclaves, with people of similar cultural background, exist in cities.
- Immigration often results from government policies and EU membership.

Over to you

Draw a spider diagram to explain how UK cultural diversity has arisen from social clustering, accessibility to cities and government policies.

You need to know:

- how and why places have changed
- how to investigate the characteristics of places.

Big idea

Past and present connections have shaped the characteristics of places.

A sense of belonging

Whether we 'belong' to a place can depend on how many people we know or are related to there.

- Morley and Dewsbury (Figure 1) were at the heart of West Yorkshire's textile (clothes) manufacturing region until the 1970s.
- Terraced back-to-back houses were typical, built by the owners of textile mills where people worked. Communities lived and worked close together.
- Family connections were important. People often married someone local, who they grew up with.
- Geographers refer to factors binding communities like these as **centripetal forces**.

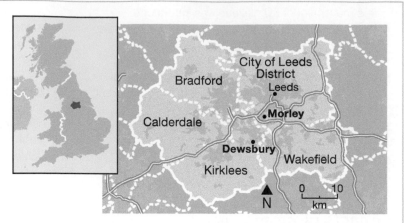

⬆ **Figure 1** *Morley and Dewsbury in West Yorkshire, which is divided into the five boroughs shown. Morley is part of Leeds city and Dewsbury is part of Kirklees*

Changing communities

Close communities like Morley and Dewsbury in the 1970s have all but disappeared.

- As Leeds has grown outwards, Morley has increasingly become part of it.
- Globalisation, bringing competition from cheaper clothing which closed textile factories and forced people to look for work elsewhere, and immigration have changed the character of many places.
- Each has forced people apart – known as **centrifugal forces**.

The impact of globalisation

By the 1980s, Morley's textile mills had all closed.

- Younger people moved away for work (**demographic change**) making Morley an ageing town by 2011.
- Textile mills (secondary employment) have been replaced by retail parks and industrial estates consisting of warehousing and home-delivery companies (tertiary).
- Most of Morley's population now commutes to Leeds and Bradford.

Immigration and cultural change

Like many industrial towns in the 1950s, Dewsbury advertised job vacancies in India and Pakistan.

⬆ **Figure 2** *Textile mills in Dewsbury, but these days the buildings are used for other purposes, such as housing and warehouses*

- It attracted mainly Pakistani Muslims, who now constitute 35% of its population.
- Immigration has changed the character of Dewsbury, with a mosque, religious community centre and sharia court.
- Dewsbury has undergone significant **cultural change**. Its close sense of community is helped by its mosque, close family ties and the town's traditional market.

Investigating places

Investigating changing places is fertile territory for fieldwork.

- Both quantitative and qualitative data help to record changes.
- You can collect either type of data, which may be **primary** or **secondary**.

Quantitative data

- **Migration surveys** collect primary data from internal and overseas migrants about relevant social processes (birthplace, migration patterns, reasons for migrating). A family member survey can focus on migration (Figure 3) and can be used to survey different parts of a locality or as a questionnaire to ask others.
- **Census profiling** uses a range of census data to identify population characteristics of an area, using social indicators such as housing, health and education. Profiles of data can be used from the National Statistics (ONS) website (Figure 4). Local councils also publish census profiles about multiple deprivation.

Qualitative data

Qualitative data include:

- **Photos** can show changes in places from past to present, and help people remember life in their own past.
- **Interviews** can record people's experiences about, for example, a sense of community.
- **Social media** (e.g. Facebook pages) feature news, photos and memories of places. Local newspapers may feature on Twitter.

Please answer the following questions about places you've lived:

1 Where were you born? (The place and the country it's in)

2 Answer all the questions below for each time you've moved home.

Place (place and country) you moved from	
How old were you when you moved here?	
What was your main reason for moving here?	
Is it a better, similar, or worse place to live than where you moved from?	

3 Please tell me the number of relatives (but not parents) living in the place where you live.

No relatives	1–4	5–9	10–19	20–29	30 and over

Figure 3 *A family member survey, which would help to calculate the extent to which the place was driven by centripetal or centrifugal forces*

	M	D
Ethnicity (%)		
White British	95.5	60.80
EU	1.8	1.80
Southern Asian (e.g. India, Pakistan)	1.5	36.50
Black or Black British	0.1	0.01
General health (%)		
General health: not good	18.0	19.50
Housing (%)		
Living in social-rented housing	12.9	23.70
Owner-occupied housing	69.9	59.30
Employment status: aged 16–74 (%)		
Employed full-time	47.0	40.20
Employed part-time	14.2	14.80
Permanently sick / disabled	3.8	6.20
Unemployed	3.9	2.75
Highest qualification level (%)		
No educational qualifications	24.9	34.20
Qualified with university degree	22.8	15.10

Figure 4 *A 2011 Census profile to compare Morley (M) and Dewsbury (D)*

Ten-second summary

- Industrial towns have changed since the 1970s.
- Changes result from globalisation, employment change and immigration.
- The characteristics of places can be measured using quantitative or qualitative data, which may be primary or secondary.

Over to you

Draw two diagrams to show how:

a centripetal forces drew together people in 1950s communities

b centrifugal forces caused by globalisation, employment change and inward migration have changed places.

You need to know:
- that the same urban places can be perceived differently by different people
- that people's perceptions depend largely on personal experiences.

Big idea

Urban places can be perceived differently by different people because of their personal experiences.

Manchester: dangerous …?

Cities are often seen as exciting places. But 19th-century cities were different (Figure 1):

- Many cities had dark underbellies, with crime such as theft and murder.
- *Mary Barton* by Elizabeth Gaskell was a murder novel set in industrial Manchester. It portrayed slums with people living in damp, unsanitary cellars dangerous to health.
- German philosopher Engels reported on England's slums in 1844, describing the conditions, poverty and filth in which people were forced to live.

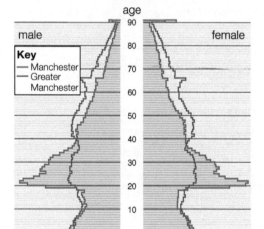

> **Figure 1**
Slum housing in Manchester at the end of the 19th century, after the worst slums had already been cleared

… or desirable?

Perceptions change! Now, 15% of Manchester's population is students, attracted by its reputation. Manchester has:

- halls of residence offering student-focused life
- thriving music, e.g. Royal Northern College of Music, a heritage of 1980s indie music and venues including The Warehouse Project
- two Premiership football teams, theatre and shopping
- a growing knowledge economy, with graduate employment in IT, science and engineering, and financial services and media (e.g. BBC Salford)
- a young population (Figure 2) – its largest cohorts are the 20–30 age band (Greater Manchester, including towns such as Salford, Oldham and Rochdale, has a very different profile).

Figure 2 *Manchester's age-sex pyramid, showing the impact of its large student population on the city*

The other side

The image of Manchester today is not all positive. Some young people are concerned about safety, and there are plenty of media stories about stabbings and urban crime to feed their fears. The reasons for poor reputations can include:

- poor environmental quality, e.g. graffiti, or run-down buildings prior to redevelopment
- racism, as people perceive migrants and other ethnicities differently

- crime rates and perceptions of crime. However, data do not always match perceptions. Although homicide is the stuff of newspapers, the rate of homicides or attempted homicide is about 6 per 100 000 people.

Most areas have mixed characteristics (Figure 3).

For more data on the five areas of Greater Manchester, see Figure 5 on page 264 of the student book.

Sale (E) and Altrincham (F), in south-west Greater Manchester, is a family area with some up-market housing and parks. It has a lower rate of crime than Fallowfield but with four times the population.

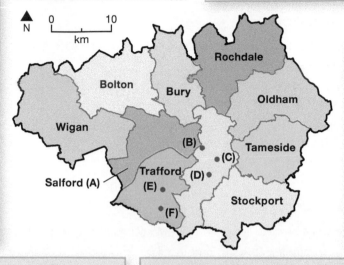
▼ **Figure 3** *Characteristics of five areas of Greater Manchester*

Salford (A) has a reputation for high crime, but perceptions are changing due to regeneration and falling crime rates. 50% of crimes are theft and anti-social behaviour. There is some up-market development, e.g. Salford Quays, but house prices are still lower than average. Population is 85.6% White British.

Fallowfield (D) is a popular residential area undergoing 'studentification' (high percentage of students). Residents complain of noise from parties. 23% of crime is anti-social behaviour and theft is 31%, typical of most student areas of UK cities.

Manchester CBD (B) consists mainly of commercial property, but housing developments are bringing people back to live here. It has a significant evening economy. 62.5% of all crimes are theft and anti-social behaviour.

Longsight (C) is an inner city area of terraced housing near the CBD. Nearly 73% of its population is classed as non-White. Gang wars caused high crime until 2009. Violent crime is falling rapidly but it still has the highest percentage of any district in the city. House prices are among the lowest in the UK.

Lived experiences and 'othering'

How different groups perceive the same urban areas depends on their experiences in those areas.

- Social media often reveals derogatory names (e.g. 'chavs') for people who live in 'other' areas. Demeaning other people in this way is known as **'othering'** and identifies people perceived to be different in some way.
- It encourages ethnic enclaves (see Section 6.3) where those who are 'othered' live close to each other for mutual support.
- With 'othering', underlying causes are left unexplained. For example, fictional Chatsworth council estate in Manchester, featured in Channel 4's *Shameless* (2004–2013) is represented as a rundown council estate full of crime not as a place of economic decline. Journalist Owen Jones claims that the series fails *'to address how characters ended up in their situation, or what impact the destruction of industry had on working-class communities in Manchester'*.

 Ten-second summary

- Cities – or different parts of them – are perceived differently by different people.
- These perceptions may vary depending on the age, gender and personal background of those involved.
- Perceptions change as places change, e.g. through regeneration or studentification.
- People's perceptions depend largely on their personal experiences.
- Whole cities or parts of cities may vary in how people perceive them. House prices may reflect these perceptions.

 Over to you

1 Draw a mind map to show how and why different parts of Manchester are perceived differently.

2 Explain why it is important for geographers to be aware of 'othering'.

You need to know:

- that the same rural places can be perceived differently by different people
- that people's perceptions depend largely on personal experiences.

Big idea

Rural places can be perceived differently by different people because of their personal experiences.

Cornwall's population boom

Cornwall is among the UK's fastest-growing counties, despite its ageing population and a 'brain drain' of young people for opportunities elsewhere.

- Inward migration brings retirees and many families looking for a rural lifestyle.
- Many incoming migrants are 'returners', who grew up in Cornwall.

Just as cities are perceived differently (see unit 6.5) by different people, people see Cornwall in different ways, with benefits and downsides.

Living in Cornwall – the upsides

Rural Cornwall has advantages as a place to live.

- **Scenery** (Figure 1) – Nowhere is far from the sea, and the coast has attracted artists, particularly to enclaves such as St Ives.
- It has a mild **climate**, especially in winter.
- **Food and drink** includes Cornish pasties, cheeses, ice cream, seafood and wines.
- **Superfast broadband** – first trialled in Cornwall, 92% of Cornish people and businesses are now connected.

Many people like Cornwall's unique isolation, which has created:

- **Identity** – There's a Cornish Nationalist Party, Mebyon Kernow, and Cornish language. Although Cornish only has 4000 speakers, they've campaigned for bi-lingual road signs.
- **Heritage** – Some of Britain's earliest settlements are in West Cornwall and the Cornish coast has a heritage of smuggling, shipwrecks and fishermen's tales.
- **Literature** – Many visit Cornwall to see settings for the *Poldark* novels (now a TV series).

🔺 *Figure 1* The classic view of Cornwall in summer – a rural idyll

Living in Cornwall – the downsides

There are downsides to Cornwall's lifestyle.

- **Geographical remoteness** from other parts of the UK. It's a long county – two hours by train from west to east – with no large cities, limiting work opportunities. It's isolated, with no motorways and slow rail links, and its only airport, at Newquay, has few flights except in summer.
- **Low incomes** – It's England's lowest income county. Tourism creates jobs that tend to be temporary, part time and low paid. Only east Cornwall has easy access to Plymouth and higher urban salaries.
- **Limited social opportunities** exist for teenagers and older people retiring away from friends and family.
- **A limited range of services** – Many village shops are under threat, and Exeter and Plymouth (both in Devon) are the only cities. Online shopping eases shopping problems. Access to healthcare can be problematic.
- **High transport costs** – Cornwall has high levels of car ownership but poor bus services, so commuting or shopping are expensive.

Just got back from holiday there. Bits of it are lovely and bits are miserable. It has gorgeous beaches but you'd need a car if you lived there.

We moved here from Manchester. Some locals are quite anti-outsiders, but others are lovely. We love the beaches but some villages are unbearably crowded with tourists in summer.

I'm from Cornwall. Looe is a lovely place to live. It's got a beach but can be boring for teenagers – not close enough to Plymouth and a bit cut off. Lots of young people leave because of the cost of living and lack of jobs.

Figure 2 *How different people perceive Cornwall as a place to live*

Cornish health services

Cornwall is one of the UK's most deprived areas, measured by:

- its shorter life expectancy
- the greater likelihood of serious illness and disability
- the proportion of adults who suffer anxiety, stay in hospital and claim health benefits.

Its deprivation stems from low wages, seasonal employment and high house prices (which are inflated by tourism and second homes).

Access to health facilities can be poor in Cornwall's remotest parts.

- Only 38% of west Cornish villages have a doctor's surgery and those that do may have surgeries only one morning a week.
- Transport is a problem – there may only be three or four buses a day in some villages. People must rely on cars or taxis for medical appointments.
- Royal Cornwall Hospital, Truro provides wide-ranging treatments and A&E. But Penzance is 26 miles away, and tourist traffic can cause delays, affecting survival rates for emergencies.

Roseland Parc – retirement village

Roseland Parc is a retirement village in Tregony, South Cornwall. These are likely to become more common as rural populations age. Roseland Parc consists of:

- sheltered flats, with a warden
- a restaurant
- nursing facilities, including for the frail and end-of-life care.

It is located in the heart of a traditional Cornish village, so it is close to shops and bus services.

It's not cheap. Flats cost £200 000, plus service charges. Care costs £4000 a month per person.

 Ten-second summary

- Rural places such as Cornwall are perceived differently by different people.
- Cornwall's advantages attract retirees and those seeking a rural lifestyle.
- Its disadvantages include its remoteness, lack of opportunity and problems in accessing services.

 Over to you

1 Create mnemonics to help you remember the list of Cornwall's advantages and disadvantages as a place in which to live and work.

2 Decide whether the positives of Cornwall outweigh the negatives.

You need to know:

- about data (quantitative) and other evidence (qualitative) that help us to know and understand how people view their living spaces
- how the media portray different places.

Big idea

How people view their living spaces varies.

How people perceive places

People's perceptions of places are mostly personal. That presents a problem – how to measure these perceptions, which are **subjective** opinions, not **objective**, based on facts?

- House prices help. A house is worth what someone will pay for it and reflects someone's desire to live there. House prices are therefore an accurate assessment of people's perceptions about a place.
- Similarly, crime data influence house prices. The national crime database (www.police.co.uk) shows whether crime correlates with house prices. Police neighbourhoods are geographical areas and show recorded crime in specific streets or even houses. These might be crimes against people or property (Figure 1).

Interpreting crime data

- It's important to consider population size. Salford (see Section 6.5) has one of the largest reported crime *totals* in Greater Manchester, but it's also got the largest population.
- It's therefore important to calculate a '**crime rate** per 1000 people'. This is done by dividing total crimes by the population, then multiplying by 1000.
- The *rate* for Salford then looks very different, with one of Manchester's lowest crime rates!

 Figure 1 Anti-social behaviour – graffiti in Hackney Wick in East London

See Figure 1 on page 270 of the student book for population, house prices and reported crime data, including crime rates, for parts of Greater Manchester in December 2015 and Figure 2 for their locations.

Collecting primary data about perceptions

Other investigations, carried out by people or organisations, can show how people feel about places.

Satisfaction with life surveys

- **Satisfaction with life surveys** can try to find out how people feel about the area in which they live.
- Questions are based on people, housing and neighbourliness, safety, environmental quality, and attitudes towards the local council, police and transport.
- This will give a list of factors that are either positive **drivers** (i.e. the things which produce high scores) or negative (i.e. things which produce low scores). Figure 2 shows an example.
- By taking a large sample, it's possible to add up the scores and express them as percentages, and then analyse which factors/ drivers produced the highest scores.

 See Figure 6 on page 272 of the student book for a satisfaction with life survey.

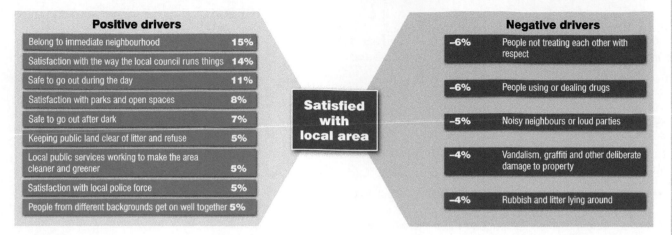

Positive drivers		Negative drivers	
Belong to immediate neighbourhood	15%	–6%	People not treating each other with respect
Satisfaction with the way the local council runs things	14%	–6%	People using or dealing drugs
Safe to go out during the day	11%	–5%	Noisy neighbours or loud parties
Satisfaction with parks and open spaces	8%	–4%	Vandalism, graffiti and other deliberate damage to property
Safe to go out after dark	7%	–4%	Rubbish and litter lying around
Keeping public land clear of litter and refuse	5%		
Local public services working to make the area cleaner and greener	5%		
Satisfaction with local police force	5%		
People from different backgrounds get on well together	5%		

Satisfied with local area

 Figure 2 *Positive and negative factors/drivers that affect whether or not people are satisfied with their local area (according to polling group IPSOS-MORI)*

Environmental Quality Surveys

Environmental Quality Surveys (EQS) can be used to measure people's feelings about the environment.

- An EQS is a bi-polar survey, scoring criteria that contribute to environmental quality.
- Scores can be collated and analysed either individually (e.g. what people think of traffic noise) or by comparing total scores from one place to another.

See Figure 8 on page 273 of the student book for an EQS.

Creating images of places

The media can represent people's feelings about places. These include:

- Music – e.g. The Specials' song 'Ghost Town' presents a negative view of Britain's inner cities in 1981.
- Photography – e.g. Figure 1, which could give a very negative image of Hackney.
- Film – e.g. *East is East* portrays multicultural communities in Salford and the challenges facing them.
- Paintings – e.g. L. S. Lowry's paintings of industrial Manchester present either a romantic or a depressed view of 'matchstick' people dwarfed against industrial buildings, depending on perception.
- Literature – e.g. Ian McKewan's novel *Saturday* is about one day in London in 2005.

 Ten-second summary

- Quantitative data, such as crime figures or house prices, reflect and influence how people view their living spaces.
- Primary data about perceptions can be collected through satisfaction with life surveys and Environmental Quality surveys.
- Different media, such as music or photography, portray different places in varying ways.

Over to you

1 Learn four key points about crime data in different parts of Manchester and the factors that influence them. You will need to look at page 270 of the student book.
2 List two media sources that reflect how different people feel about a place you have studied.

You need to know:
- how internal and international migration affect culture and society
- why some international migrants choose to live in rural areas.

Big idea
Culture and society in the UK are increasingly diverse.

Population change

London is now bigger and growing faster than at any time. Part of the reason is internal migration.

- Many graduates elsewhere in the UK find it difficult to get well-paid work, so they move to London.
- London and the South East receive more internal migrants than anywhere else.

The process of population change consists of inflows and outflows of people (Figure 1).

- In 2013, UK net population increased by over 400 000.
- 56% came from **natural increase** (i.e. births minus deaths).
- 44% came from **net migration** (i.e. immigration minus emigration).

▼ **Figure 1** *The UK's population dynamics in 2013, shown as a system with five* **variables**

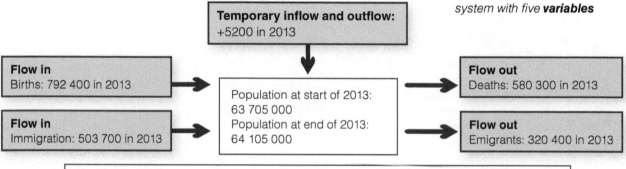

Temporary inflow and outflow: +5200 in 2013

Flow in
Births: 792 400 in 2013

Flow in
Immigration: 503 700 in 2013

Population at start of 2013: 63 705 000
Population at end of 2013: 64 105 000

Flow out
Deaths: 580 300 in 2013

Flow out
Emigrants: 320 400 in 2013

Population change = (births minus deaths) +/- (immigration minus emigration) +/- (temporary inflow minus temporary outflow)

London's population dynamics

In 2013, London's population increase was over 40% of the UK total. It had the UK's largest natural increase (82 900, 40% of the UK's increase) and net international migration (79 500, 43% of the UK's net total). However, an even greater **outflow** took place from London.

Moving away

London's outflow in 2013 was higher than anywhere – 55 000 (Figure 2).

- It consisted mainly of families with children (who moved to the South East) and retirees (who moved mainly to the South West), helped by London's high property prices.
- As a result, London now has increasing numbers of young professionals and ethnic communities.
- In 2011, Newham (East London) became the first London borough where White British people were a minority.

London's outflow has occurred since the 1960s.

- Geographers used to speak of 'white flight' – middle-class white families leaving cities for the suburbs.
- As they left, better-off migrants moved from inner cities to the places vacated by the middle classes.
- The same process occurs when British Asian families from Southall move, e.g. to Iver in Berkshire.

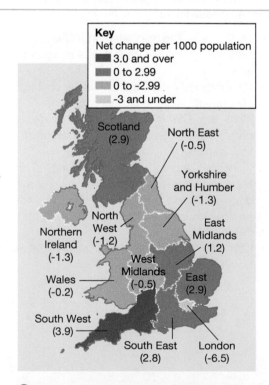

Key
Net change per 1000 population
- 3.0 and over
- 0 to 2.99
- 0 to -2.99
- -3 and under

Scotland (2.9)
North East (-0.5)
Yorkshire and Humber (-1.3)
North West (-1.2)
Northern Ireland (-1.3)
East Midlands (1.2)
West Midlands (-0.5)
Wales (-0.2)
East (2.9)
South West (3.9)
South East (2.8)
London (-6.5)

▲ **Figure 2** *Net change per 1000 population as a result of internal migration within each of the UK's standard economic regions*

UK immigration

International migration has changed UK culture and society.

- Since 1945, many migrants have come from Europe and southern Asia.
- Those born beyond Europe come mainly from former colonies, e.g. India, Pakistan, West Indies.
- There have been two phases, each resulting from government policies.

Post-colonial migration

The British government promoted immigration from former colonies from the 1940s to the 1960s.

- It gave migrants the right to settle in the UK and sponsored them to fill job vacancies (e.g. in transport).
- By 2014, they formed the majority of non-UK-born residents.
- Many of their families still live in areas of deprivation.

The impacts of globalisation

Since the 1990s, the UK population has risen rapidly because:

- TNCs encourage immigration, e.g. from the USA and Australia, to work in London's knowledge economy.
- The 1992 'Maastricht Agreement' allowed free movement of workers between EU member states, with increased economic migration after 2004.
- Younger migrants balance an ageing UK workforce and provide unskilled as well as skilled work.

In addition, many immigrants from countries such as India and Pakistan are joined by family members as part of their right to settle.

For the numbers of non-UK born, EU born and non-EU born residents in the UK economic regions in 2014, see Figure 5 on page 276 of the student book.

Migrants in rural areas

Rural areas are affected by EU migration, e.g. vegetable pickers in Lincolnshire (Figure 3). Their arrival has had a huge impact.

- The overseas-born population of Boston, Lincolnshire, increased by 470% from 2001 to 2011. Some have welcomed migrants, but others have not, believing migrants undercut wage rates.
- In 2015, the *Financial Times* reported Latvian pickers were paid below the farm minimum wage and excessive costs for transport and accommodation had also been deducted from their wages.

▲ **Figure 3** *Vegetable pickers in Lincolnshire. Why are migrants more likely to head for urban rather than rural areas when they move to the UK?*

 Ten-second summary

- Internal migration within the UK has created uneven population and cultural patterns, as has international migration from former colonies and the EU.
- Some international migrants choose to live in rural areas.

 Over to you

1 Draw an annotated diagram to show why London is a more diverse city with a younger population than anywhere else in the UK.
2 Draw a mind map to show how government policies influence flows of people to the UK.

You need to know:

- how segregation reflects cultural, economic and social variation
- how people's perceptions of places change over time.

Big idea

Levels of segregation reflect cultural, economic and social variation, which change over time.

The super-rich

London now attracts the super-rich, including ministers from oil states, rulers of overseas countries and **oligarchs** (members of a wealthy, powerful group who control a country or organisation).

In 2013, about 100 Russian billionaires bought London property; more than in any other country. Many did so to escape economic sanctions imposed against Russia after the 2014 Crimean crisis and to protect assets by purchasing property in safe locations. Most live in **segregation** in self-contained, security-guarded enclaves in Central London.

For investors, London's attractions are:

- Poperty, as it is a good investment – average prices rise by 10% annually

- The British pound holds its value more consistently than Russia's rouble; the Russian economy is volatile and dependent on oil and gas.
- Russians use London to raise capital for their companies.
- The UK grants the super-rich three-year visas if they invest in government bonds. After two years, residency costs £10 million. Over 800 visas were granted to Russians and Chinese between 2008 and 2013.

Other reasons for choosing London include:

- Education in private schools for children – in 2013, over 8% of non-British pupils in independent schools were Russian.
- Low crime rates – the UK had fewer than 600 murders in 2014, compared to Russia's 16 000.

London – ethnicity and culture

Ethnic segregation is common in many cities. It is apparent in the wide range of distinctive cultures and religions in London (Figure 1).

- Renting property makes it easy for migrants to settle. Obtaining employment then makes such communities permanent (e.g. Sikh communities in Southall).
- West Indian migrants were recruited to drive London's buses, so many settled near bus garages, e.g. Brixton. Many descendants still live in the same areas.
- Once established, the growth of shops, places of worship and leisure facilities reinforce culture and make ethnic communities more permanent.

Figure 1 *A greengrocer's shop in Brixton, South London – one of the indicators of local ethnicity in inner cities*

London's Greek Cypriot community

Most of London's Greek Cypriot community arrived in the 1960s and 1970s. Including migrants from mainland Greece, there are now 180 000 Greek speakers in London. The Greek Orthodox Church plays a significant role in language teaching, ensuring that their **assimilation** into the UK also retains their **cultural distinctiveness**.

The Greek Cypriot community lives in North-West London (Figure 2). It has established:

- several places of worship, with Greek Orthodox churches, shops and other businesses, and a public cemetery
- a Greek Cypriot newspaper and a London radio station.

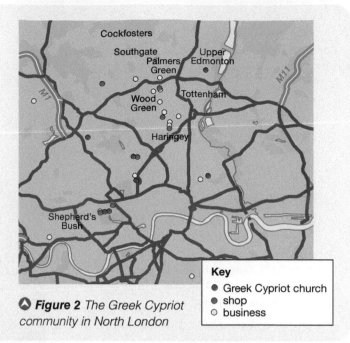

Figure 2 *The Greek Cypriot community in North London*

Key
- ● Greek Cypriot church
- ● shop
- ○ business

Measuring cultural diversity

Many cities and countries have diverse populations. Inner London is Europe's most diverse place. **Cultural fractionization,** which measures the diversity of countries, can be shown on a map by:

- measuring people's attitudes towards religion, democracy and the law
- using these to create an index between 1 (total diversity) and 0 (no diversity).

The global average is 0:53.

For a map showing global **cultural fractionization**, see Figure 6 on page 280 of the student book. Darker-shaded countries are more diverse than lighter.

How people experience change

Inward migration changes the character of places and communities, together with people's views about them. Former residents may see a place change and cease to identify with it once a new identity emerges.

- Brick Lane in East London has seen waves of migrants, including Protestant Huguenots in the 17th century and Jewish refugees from Russia and Germany in the 20th century.
- Over time, London's Jewish community moved to Golders Green, and Brick Lane became home to a Bangladeshi community with its restaurants, shops and mosques.
- However, Brick Lane borders the City and the community is threatened by **gentrification**, caused by higher property prices paid by City workers.

 Ten-second summary

- International migrants tend to live in distinctive places.
- Diversity in urban areas reflects the ethnicity and culture of those living there.
- Experiences of living spaces change over generations as communities evolve.

 Over to you

1 Summarise the factors that attract the super-rich to London.
2 In a table, list the economic and cultural contributions of ethnic communities to London.

You need to know:
- how changes to land use can create challenges and opportunities
- why changes to diverse places can lead to tension and conflict.

Big idea
Changes to diverse places can lead to tension and conflict.

Global changes to local places

London Gateway, the UK's newest port, lies 30 km from Central London. It takes the world's largest container ships.

- Before 1981, container shipping led to the closure of London's original port because the Thames was too shallow.
- The causes were globalisation and the shift of manufacturing to Asia, which changed the economics of shipping.

The last of London's original docks closed in 1981. Their closure was devastating.

- Economically, the area was in a spiral of decline. Between 1978 and 1983, 12 000 jobs were lost, creating 60% unemployment in adult men.
- Socially, the early 1980s were marked by unrest in London's more deprived districts, e.g. Brixton. Race relations with the police raised tensions and unemployment tested this further.
- Environmentally, 21 km² of riverside property was left derelict or run-down (Figure 1).
- Demographically, London Docklands lost 100 000 people between 1971 and 1981, as people left to seek work.

▲ **Figure 1** *Derelict buildings, like this former flour mill, lined the Thames downstream from Central London in the 1980s*

Regenerating London Docklands

The closure of London's docks created huge potential for regeneration.

- Planning it went to a government agency, the LDDC (London Docklands Development Corporation).
- It aimed to encourage investment from developers in a process called **market-led regeneration**, leaving the private sector to decide Docklands' future.
- The LDDC focused on economic growth, infrastructure and housing.

- Local councils had no say in LDDC plans. There were tensions between what they wanted (e.g. affordable housing) and the plans approved by the LDDC. The Docklands regeneration was **top-down**, ignoring community visions for East London.
- LDDC's flagship was Canary Wharf, consisting of offices to attract quaternary employment. 100 000 people commute there daily, with average annual salaries of over £100 000. Employment has grown but, in 2012, 27% of Newham's working population earned under £7 per hour – the highest percentage of any London borough.

Gentrification

Changes to Docklands have created winners and losers, and have transformed the population of the area.

- Regeneration increased housing supply, but of luxury housing in riverside locations, which has led to **gentrification** (Figure 2) and migration of the middle classes into East London.
- A division has emerged between new and old 'Eastenders'. New young professionals drink and eat in renovated 'foodie' pubs and restaurants, while the traditional East End pub struggles to survive.

Changes to social housing

Before regeneration, most East London housing was rented from local councils. Two major changes occurred:

 Figure 2 Canary Wharf after redevelopment

- The Conservative government introduced the **Right to Buy** scheme, giving council tenants the right to buy the houses they lived in. This reduced the supply of social housing so that, by 2016, 20 000 people were on Tower Hamlets' housing waiting list.
- East End residents who bought council houses in the 1980s have now sold them and left to retire on the Essex coast. They've been replaced by a younger generation. In 2011, the average age in Docklands was 31.

Traditional East End communities have been broken up.

- Few people know their neighbours.
- People tend not to stay for long, moving from one rental property to another.
- Essex now has more traditional Eastenders than Newham or Tower Hamlets.

Migrant experiences in East London

Increased immigration since 2000 has made East London the UK's most ethnically diverse area.

- Some migrants experience isolation and hostility from those who claim that migrants threaten 'British' culture.
- Hate crimes against Muslims in London have risen, especially against women and those wearing a face veil.
- East European migrants experience less aggression, but may have little interaction with other Eastenders.
- However, London is generally perceived as a liberal and open-minded city. It voted overwhelmingly against Brexit and in 2016 elected Sadiq Kahn, its first Muslim Mayor, reflecting its tolerance towards a wide range of ethnicities.

Ten-second summary

- Changes to Docklands have created challenges and opportunities.
- Changes to Docklands have brought benefits to some but provoked hostility from others.
- Tensions can arise over diversity of living spaces, between long-term residents and recent in-migrants.

Over to you

Draw a spider diagram to show how the following have affected local people:

- the closure of London's docks
- London Docklands regeneration
- the Right to Buy
- out-migration of Eastenders
- gentrification
- large-scale immigration.

You need to know:
- how economic and social inequalities can be measured
- how the assimilation of different cultures can be measured.

Big idea
Cultural and demographic issues can be measured.

Cities as unequal spaces

Inequalities exist in all urban areas: the larger the city, the greater the inequality (Figure 1). Although London is far from being the world's most unequal city, inequality is growing. A 2011 report showed that:

- Of all UK economic regions, London had the highest proportion of households in the top 10% of incomes *and* the highest proportion in the bottom 10%.
- Inequality in London grew during the 1997–2008 economic boom. The incomes of its wealthiest grew, but poverty levels were unchanged.

As London has become more diverse, so inequality has increased, because of economic activity among different ethnic groups, low pay and housing costs.

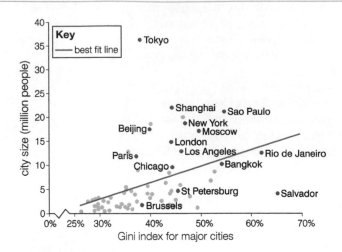

▲ *Figure 1* The relationship between city size and inequality, as measured by the **Gini index**

Economic inactivity

Economic inactivity (being without work), whether through unemployment or disability, is a major cause of poverty.

- Though most migrants to the UK are economic, employment levels vary by ethnicity.
- In 2011, Bangladeshi and Pakistani women had highest economic inactivity among adults, because of their traditional family roles.

Housing costs

London incomes are the UK's highest, but poverty has increased due to housing costs.

- UK inflation rose by 73% between 1995 and 2015, but house prices rose by up to 1000%!
- Londoners spend up to 60% of their income on housing, compared with the UK average of 25%.
- Many can't afford to live in large areas of London, worsened by government cuts to housing benefit since 2010.
- One of London's key problems is attracting workers into essential services such as education.

Low pay

Low pay is a major cause of poverty (Figure 2). In 2011:

- 37% of employed men worked in low-pay jobs, the highest rates being among Pakistani and Black African men.
- 59% of employed women were in low-skill jobs (e.g. NHS, social care). Bangladeshi and Caribbean women fared worst.
- 40% of ethnic minority people live in low-income households, twice the rate of White British people.
- 10% of White British employees are in low-pay jobs. With Bangladeshis, it's 65%.

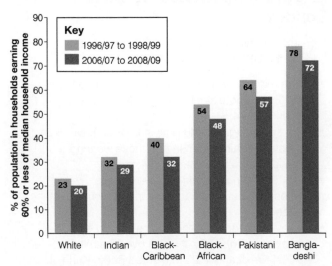

▷ *Figure 2* The percentage of people earning 60% or less of median household income, after deduction of housing costs

Tackling inequality

Regeneration should increase wealth – jobs are created in building and inward migration brings wealth to the area. East London has high levels of deprivation, so *should* benefit from regeneration (e.g. Docklands, the 2012 Olympic Games). But has it?

- **Barking and Dagenham** was one of London's biggest manufacturing boroughs until Ford (cars) and Sanofi (pharmaceuticals) closed many production facilities. Regeneration is now focused on housing. Incomes rose by 6% in 2013.
- **Tower Hamlets** is one of London's poorest boroughs, though Canary Wharf means it has London's highest salaries. But those who work in Canary Wharf live outside the area, so regeneration created few local jobs. Incomes rose by 5% in 2013 but not for local people.

- **Newham** hosted the 2012 Olympic Games and contains Westfield Shopping Centre, generating 10 000 jobs. However, the Games employment was temporary and retail jobs tend to be part-time and low paid. Incomes actually fell by 6% in 2013.

Deprivation in London reduced between 2010 and 2015 (Figure 3). Is this due to regeneration? Data should be treated carefully:

- In 2010, the UK was recovering from the global banking crisis. Economic improvements meant that deprivation fell as unemployment fell.
- Many health indicators were not linked to regeneration but to targets in the NHS, e.g. in diagnosing and treating cancer. These would count as reductions in deprivation but would not be due to regeneration.

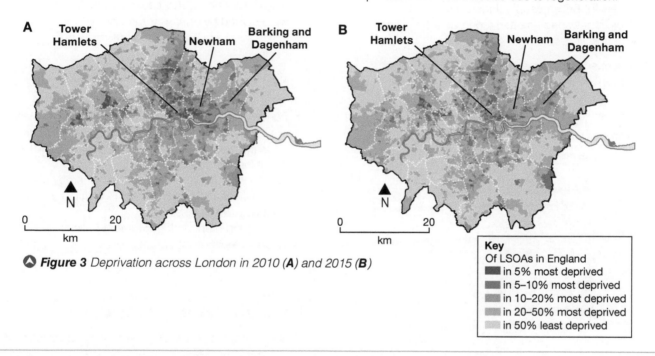

Key
Of LSOAs in England
■ in 5% most deprived
■ in 5–10% most deprived
■ in 10–20% most deprived
■ in 20–50% most deprived
■ in 50% least deprived

 Figure 3 *Deprivation across London in 2010 (**A**) and 2015 (**B**)*

Assimilation and engagement

London is diverse, but the participation and assimilation of ethnic groups varies.

- Voter turnout varies during elections and the poorest are least likely to vote. However, diverse ethnicity *increases* the likelihood of voting, perhaps because people experience prejudice or exploitation.
- Despite falling crime in London, hate crimes are common, especially against London's Muslim community. Now local Muslim communities and mosques hold open days to help people understand Muslim life.

 Ten-second summary

- Inequality can be assessed economically using income and employment.
- Inequality can be assessed socially by measuring reductions in deprivation.
- Assimilation of different cultures can be measured by indicators such as voter turnout and reductions in hate crime.

Over to you

Using highlighters, identify how different factors contribute to poverty in London.

You need to know:

- how different national and local strategies can be used to resolve multicultural issues
- that different stakeholders (players) have different criteria for assessing change in diverse urban communities.

Big idea

Urban stakeholders (players) have different criteria for assessing change in diverse communities.

Diversity and changing communities

During the 2012 Games, London was praised for its integration of migrants – evidence that its **multicultural** society was working well. Yet, outside London, immigration is contentious.

- Immigration features in the top five most important issues for British people (Figure 1).
- Tabloid newspapers often report about immigration with hostility.

Yet experiences of racism in Britain are infrequent (Figure 2). How can integration like this be achieved?

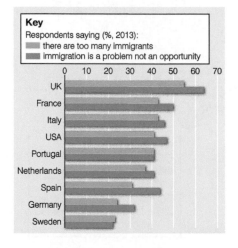

Key
Respondents saying (%, 2013):
■ there are too many immigrants
■ immigration is a problem not an opportunity

○ **Figure 1** *How people perceive immigration in the UK compared to other countries*

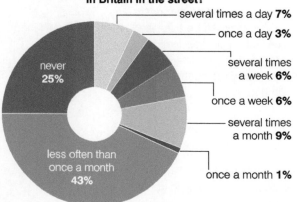

Have you ever personally encountered racism in Britain in the street?

- several times a day **7%**
- once a day **3%**
- several times a week **6%**
- once a week **6%**
- several times a month **9%**
- once a month **1%**
- less often than once a month **43%**
- never **25%**

○ **Figure 2** *Experiences of racism on the streets of Britain*

National strategies towards multicultural issues

British attitudes towards multiculturalism have changed since the late 1960s, when rising tensions between different ethnic groups created **centrifugal forces**, driving people apart. However:

- Early 1980s riots in cities such as London and Leeds were only partly caused by racism. Others were started by right-wing groups.
- Often, riots were linked to protests about police treatment.

Since then, different governments have adopted different policies about multiculturalism.

- Early immigrants in the 1950s were **segregated** into enclaves, with little understanding between different communities. London's annual Notting Hill Carnival sought to bring people together.
- Later policies sought **assimilation**, whereby different ethnicities would integrate into UK culture by adopting 'Britishness'.
- Now governments adopt **pluralist** policies, which value the cultural contribution of every ethnic group.

Local strategies in Slough

Slough (population 140 000) is like many towns located 20–40 miles from London. It lies along the M4 corridor, west of London beyond Heathrow (Figure 3).

- Slough shows how ethnic tensions can be avoided. In 2011, 18 ethnic groupings lived there, with 34.5% White British population (UK average is 80.5%).
- Over 75% claimed to be British, even though only 60% were born in the UK. So many have adopted British nationality.

There are no enclaves but long-established migrant communities instead.

- In the 1920s and 1930s, Welsh and Scots moved to Slough, followed by Polish (1940s) and then Indian migrants (1950s).
- There have been some ethnic tensions, e.g. after the murder of Stephen Lawrence in South London.

Key players in regenerating Slough

Slough Borough Council

The Borough Council is a key **player** and decides priorities for Slough, e.g. how land should be used (housing, employment space, recreation). It has identified sites for regeneration, focusing on:

- housing (the biggest need)
- a cultural learning centre (the Curve), with library, adult education facility, cafe and performance centre
- a sports stadium (at Arbour Park), with football and athletics facilities, games areas, clubhouse and accommodation block.

Community groups

Aik Saath, a local charity, tried to reduce tensions between gangs of young Muslims, Hindus and Sikhs in the 1990s. It focuses on:

- young people, aiming to increase **centripetal forces** to bring people together
- working with teachers on themes such as extremism and anti-racism
- training in schools and with the police about how young people, teachers and youth workers can deal with sensitive issues.

Now, many see Slough as a multicultural success story.

Economic stakeholders

Slough is prosperous.

- Its proximity to Heathrow (Figure 3) has led TNCs to establish offices, and people have integrated to workplaces. The Borough Council has attracted many companies to work with it, including a commercial partnership to regenerate the town.
- However, housing is in short supply, worsened by cuts to housing benefit, forcing people out of London.
- Aspire Southall is a partnership between companies in Slough and Slough Borough Council, developing employability skills among young people. It has a training centre managed by local companies.

Environmental stakeholders

Slough Borough Council is responsible for community health, improving the environment (e.g. air quality), transport and safety. Air quality is hard to manage so close to the M4 and Heathrow.

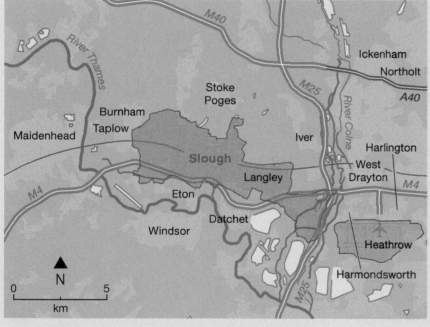

Figure 3 The location of Slough

Over to you

1 Learn two points shown by Figure 1 and two points shown by Figure 2.
2 Draw two diagrams to show:
   a centrifugal forces driving people apart in the UK
   b how Aik Saath and other players have tried to develop centripetal forces in Slough.

You need to know:
- how different strategies can be used to help regenerate rural areas
- how different stakeholders (players) have different criteria for assessing change in diverse rural communities.

Big idea
Rural stakeholders (players) have different criteria for assessing change in diverse rural communities.

Diversity in Cornwall

The South West is one of the UK's least ethnically diverse regions. 95% of the population is White British. Increasingly, migrant workers have widened the range of ethnic backgrounds in Cornwall. Most migrants have a positive effect on Cornwall's economy, but may earn only £2 per hour after deductions.

Cornwall – image versus reality

Cornwall is the UK's top tourist destination. But the reality for people living there is different.

- The county's 'old economy' consisted of primary sector employment in farming, fishing, tin mining and china clay quarrying (Figure 1). These jobs were year-round and permanent.
- Supermarkets have driven down the milk price paid to dairy farmers, so many farmers have now switched to arable or vegetable farming. With Cornwall's hilly slopes and wet winters, there has been a big increase in soil erosion.
- As the primary sector has declined, the **post-production countryside** economy focuses on tourism. These jobs are mostly seasonal, low-paid and part-time.
- Cornwall needs a year-round economy for adult employment. The lack of opportunity for young people causes a 'brain drain'.
- Cornwall has England's lowest full-time average annual earnings of £25 155 (77% of the UK average).

⬥ **Figure 1** *Wasteland produced by Cornwall's china clay industry*

National and local strategies for rural areas

Until 2007, rural areas like Cornwall could attract investment from the UK Government and the EU (known as Objective One). However, the Government cut funding for rural investment after 2010 and it is now much harder for rural counties such as Cornwall to attract private investment without some kind of Government support.

Now, **Enterprise Zones** have been designed to offer regional aid to stimulate investment. In 2015, there were 44 in the UK (Figure 2). These offer incentives for investors, including:

- business tax discounts of up to £160 000 a year.
- no need for planning permission (except safety building regulations)
- superfast broadband
- tax allowances against the costs of buildings and training.

▶ **Figure 2** *Areas (in red) qualify for* **Regional Aid** *in the UK*

Case studies of investment in Cornwall

The Eden Project

A tourist attraction, it consists of two large 'biomes', an education centre and a hostel for residential groups.

Developed from an old china-clay quarry, it cost £140 million. 75% came from the Lottery, the government and EU Objective One funding.

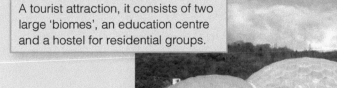

It has generated £2 billion for the Cornish economy, attracted 15 million visitors and employed 650 people.

Figure 3 *The Eden Project*

Superfast broadband

By 2016, over 95% of Cornwall had fibre broadband and it has the world's largest rural fibre network. It was granted £53.5 million from the EU and £78.5 million from BT, benefiting tourism, knowledge-economy companies and home workers. One evaluation showed that 2000 jobs had been created, with an annual economic impact of around £200 million.

Newquay Aerohub

In 2014, Cornwall Council obtained Enterprise Zone status for Aerohub Business Park near Newquay Airport in a partnership with private investors aiming to diversify Cornwall's economy.

- Its 'brand' is its location, attracting aviation companies.
- It aimed to create 700 skilled permanent jobs by 2015.

By 2015, businesses there included:

- **Aircraft industries** – flight training for pilots, engineers, aircraft maintenance, Coastal Search and Rescue, offshore operations for the RAF and Navy, Cornwall's Air Ambulance, and an aviation museum.
- **Others** – e.g. wind turbine maintenance, a development centre for the Bloodhound Super Sonic Car and a modular buildings manufacturer.

Impressive, but generating only 450 jobs, few of which were 'new'. Many jobs were transferred from elsewhere in Cornwall or were 'displaced' during privatisation of government services.

Key players in managing change in Cornwall

- **The EU** – Still funds infrastructure – until Brexit occurs.
- **Cornwall Council** – There has been no start-up funding since 2010. Instead, the council offers incentives as part of its Enterprise Zone.
- **Stakeholders in the local economy** – The biggest industry is tourism. It wants better infrastructure (roads, rail and air travel).
- **Environmental stakeholders** – Many are concerned about the waste created by clay extraction and soil erosion from arable land.
- **Stakeholders in people** – Education, e.g. Combined Universities in Cornwall and Cornwall FE Colleges.

 Ten-second summary

- Regeneration has suffered from government cuts and withdrawal of EU funding.
- A range of regeneration projects exists in Cornwall.
- Different criteria can be used to judge the success of regeneration.

 Over to you

Assess each of the regeneration (investment) projects mentioned here, based on:

- how well it suits Cornwall's environment
- its economic, social, environmental costs and benefits
- employment
- whether it develops a 'knowledge economy' and helps to prevent 'brain drain'.

Glossary

This glossary includes key words and terms related to your Edexcel A Level Geography specification.

A

A T Kearney Index – measures how globalised a country has become

ablation zone – where outputs from a glacier exceed inputs

abrasion – the grinding away of bedrock by fragments of rock which may be incorporated in ice. Also known as *corrasion

accumulation zone – where inputs to a glacier exceed outputs

adaptation – strategies designed to prepare for and reduce the impacts of events

Arctic amplification – the phenomenon where the Arctic region is warming twice as fast as the global average

asthenosphere – the part of the mantle, below the *lithosphere, where the rock is semi-molten

attrition – the gradual wearing down of rock particles by impact and *abrasion, leading to a reduced particle size and rounder, smoother stones

B

basal sliding (or slip) – where pressure and friction create meltwater at the base of the glacier, which lubricates the glacier's movement

Benioff zone – the area where friction is created between colliding tectonic plates, resulting in intermediate and deep earthquakes

C

Central Business District (CBD) – the area of a city which is usually the hub of finance, services and retail

centrifugal forces – forces which push people apart, for example changes in employment

centripetal forces – forces which draw people together, for example a strong sense of community

cirque glaciers – see *corrie glaciers

climatic climax community – the final community of species that will be adjusted to the climatic conditions of an area

closed system – there are no inputs or outputs of matter from an external source

compressional flow – when reductions in gradient force a glacier to slow down, causing it to 'pile up' and thicken

concordant coast – where bands of more-resistant and less-resistant rock run parallel to the coast

convection currents – hot, liquid magma currents moving in the *asthenosphere

corrasion – see *abrasion

corrie glaciers – small glaciers occupying hollows on mountains which may feed into *valley glaciers

corrosion – the breaking down of rock by chemical action, often involving the dissolving of alkaline rock by weak acids in seawater

cost-benefit analysis – a process by which the financial, social and environmental costs are weighed up against the benefits of a proposal in terms of social outcomes as well as in terms of profit and loss

crustal fracturing – when energy released during an earthquake causes the Earth's crust to crack

cryosphere – the frozen part of the Earth's hydrological system

cultural diffusion – the spread of one culture to another by various means

cultural erosion – the changing and loss of culture in an area, such as the loss of language and traditional food

cultural fractionization – measures how diverse countries are, by measuring people's attitudes towards, for example, religion, democracy and the law

D

Dalmatian coasts – a type of *concordant coastline formed as a result of a rise in sea level when valleys flooded leaving the tops of the ridges above the surface of the sea as offshore islands

deindustrialisation – the decline of manufacturing industry in an area

dependency ratio – the ratio of dependents (aged under 15 or over 64) to the working-age population (aged 15-64), published as the proportion of dependents per 100 working-age population

deregulation – the reduction in rules which means that any foreign business can set up in the UK

discordant coasts – the geology alternates between bands of more-resistant and less-resistant rock, which run at right angles to the coast

dormitory suburbs – residential areas which are primarily homes for commuters

drift-aligned – where sediment is transferred along the coast by *longshore drift producing a pattern of sediment size and roundness which varies between one location on a beach and another

dynamic equilibrium – where landforms and processes are in a state of balance

E

ecological footprint – a measure of the land area and water reserves that a population needs in order to produce what it consumes (and absorb the waste it generates), using current technology

economies of scale – the ability to reduce costs proportionately by increasing the scale of production

emergent coastline – when a fall in sea level exposes land previously covered by the sea

Enterprise Zones – small areas which offer incentives to attract companies, such as tax discounts and a reduction in planning permission requirements

entrainment – the process by which surface sediment is incorporated into a fluid flow (e.g. air, water or ice) as part of the process of erosion

Environmental Impact Assessment (EIA) – a quantitative means of estimating the environmental changes arising from a proposal

epicentre – the point on the Earth's surface directly above the *focus of an earthquake

equilibrium line – the boundary between the *accumulation zone and the *ablation zone

ethnic enclaves – concentrations of particular communities in an area, such as a high concentration of Asian or Asian British residents with a Pakistani background in East London

eustatic change – when the sea level itself rises or falls, partly as a result of the growth and decay of ice sheets

Export Processing Zones – the term now used in China for *Special Economic Zones (SEZ)

extensional flow – when an increase in gradient causes the ice to flow faster and it become 'stretched' and thinner

F

focus – the point inside the Earth's crust from which the pressure is released when an earthquake occurs

Foreign Direct Investment (FDI) – investment made by an overseas company or organisation into a company or organisation based in another country

G

Gini coefficient/Gini index – fundamentally, these measure the same thing: inequality. They measure how far wealth distribution within a country deviates from perfect equality. The Gini index is shown as a percentage, and the Gini coefficient as a value between 0 and 1 (i.e. the Gini index divided by 100). The larger the number, the greater the concentration of wealth will be amongst only a few. An index/coefficient rating of 0 = perfect equality, and a rating of 100 or 1 = perfect inequality

glacial outburst – when a huge amount of meltwater, that was previously trapped either beneath the ice or as surface lakes, eventually bursts

global homogenisation – the idea that everywhere is becoming the same

glocalisation – when a company re-styles its products to suit local tastes

Gross Domestic Product (GDP) – the same as *Gross National Income, but excluding foreign earnings

Gross National Income (GNI) – the value of goods and services earned by a country (including overseas earnings), formerly known as Gross National Product (GNP)

H

Haff coast – a *concordant coastline which consists of long spits of sands and lagoons

hazard-management cycle – a theoretical model of hazard management as a continuous four-stage cycle involving mitigation, preparation, response and recovery

hazard-response curve – see *Park model

hot spot – points within the middle of a tectonic plate where plumes of hot magma rise and erupt

hub cities – see *world cities

Human Development Index (HDI) – a measure of development which takes into account life expectancy, education and *GDP for every country and converts them into a value between 0 and 1

hydrometeorological hazards – natural hazards caused by climate processes (including droughts, floods, hurricanes and storms)

hyper-urbanisation – rapid *urbanisation

I

imbrication – where glacial deposits are orientated (or aligned), overlapping each other like toppled dominoes

Index of Multiple Deprivation – an overall measure of deprivation which incorporates income, employment, education, health, crime, barriers to housing and services, and living environment

Integrated Coastal Zone Management (ICZM) – a strategy designed to manage complete sections of the coast, rather than individual towns or villages, by bringing together all of those involved in the development, management and use of the coast

Inter-Governmental Organisations – organisations which comprise of two or more countries working together. Examples include the EU and the United Nations

International Monetary Fund (IMF) – a global organisation whose primary role is to maintain international financial stability

interstadials – short-lived warmer periods within a major glacial, associated with ice retreat

intra-plate earthquakes – earthquakes which occur far from plate margins

isostatic change – when the land rises or falls, relative to the sea, often in response to the melting or accumulation of glacial ice

Glossary

K

knowledge economy – see *new economy. Also associated with the quaternary sector of industry which provides highly specialised jobs that use expertise in fields such as finance, law and IT

KOF Index – measures how globalised a country has become by taking into account international interactions

L

L waves – the slowest seismic waves, which focus all their energy on the Earth's surface

land-use zoning – a process by which local government regulates how land in a community may be used

liquefaction – when the violent shaking during an earthquake causes surface rocks to lose strength and become more liquid than solid

lithosphere – the solid layer, made from the crust and upper mantle, from which tectonic plates are formed

littoral zone – another name for the coastal zone: the boundary between land and sea which stretches out to sea and onto the shore

longshore drift – the movement of sand and shingle along the coast

Lorenz curve – a graph to show and measure any inequality by comparing it to a line of perfect equality

M

market-led regeneration – the improvement of an area which is driven by the potential needs and wants of customers

mass movement – the downward movement of material under the influence of gravity. It includes a wide range of processes such as rockfalls, landslides and *solifluction

Milankovitch cycles – three interacting astronomical cycles in the Earth's orbit around the Sun, believed to affect long-term climatic change

mitigation – action to reduce the impacts of an event

multiple-hazard zone – an area that is at risk from multiple natural hazards such as hurricanes and earthquakes

multiplier effect – when success in one business creates further wealth and spending, boosting the economic development of the local economy as a whole

N

negative feedback – the regulation and reduction of a natural process

negative multiplier – a downward spiral or cycle, where economic conditions produce less spending and less incentive for businesses to invest (therefore reducing opportunities)

neo-liberalism – a belief in the free flows of people, capital, finance and resources. Under neo-liberalism, State interventions in the economy are minimized, while the obligations of the State to provide for the welfare of its citizens are diminished

new economy – where *GDP is earned more through expertise and creativity in services such as finance and media than from the manufacture of goods. Also known as the *knowledge economy

nivation – a range of processes associated with patches of snow

Non-Governmental Organisation (NGO) – a non-profit organisation created by private organisations or people with no participation or representation by any government

O

off-shoring – when a company does work overseas, either itself or using another company

open system – where energy and matter can be lost to and gained from an external source

outsourcing – when work is contracted out to another company (known as *off-shoring when that company is overseas)

P

P waves – the fastest seismic waves which travel through both solids and liquids

palaeomagnetism – the study of past changes in the Earth's magnetic field

paleo-environments – fossil or geologically past environments

Park model (hazard-response curve) – shows how a country or region might respond after a hazard event

periglacial – areas at the edge of permanent ice, characterised by *permafrost and a *tundra climate

permafrost – where a layer of soil, sediment or rock below the ground surface remains almost permanently frozen

pioneer species – the first colonising plants which begin the process of plant succession

polders – land, often reclaimed from the sea, enclosed by embankments

positive feedback – enhances and speeds up processes, promoting rapid change

post-industrial economy – see *new economy

post-production countryside – the situation in rural areas when the 'old economy', consisting mainly of primary sector jobs, has declined

pressure and release (PAR) model – a tool used to work out how vulnerable a country is to hazards

pressure melting – lower down a glacier, where pressure is higher, ice may melt at temperatures below 0°C

pressure melting point – the temperature at which ice is on the verge of melting

Purchasing Power Parity (PPP) – relates average earnings to local prices and what they will buy. This is the spending power within a country, and reflects the local cost of living

Q

quotas – a fixed level indicating the maximum amount of imported goods or persons which a state will allow in

R

re-imaging – how the image of a place is changed and portrayed in the media

regelation – melting and freezing of ice, caused by changes in pressure

remittance payments – income sent home by individuals working elsewhere (usually abroad but can be in urban areas)

rural-urban continuum – the spectrum which moves from a large city or conurbation to countryside areas

S

S waves – seismic waves which only travel through solids and move with a sideways motion

sediment budget – the amount of sediment available within a *sediment cell

sediment cells – a length of coastline and its associated nearshore area within which the movement of coarse sediment (sand and shingle) is largely self contained. There are 11 sediment cells around England and Wales some of which can be divided into sub-cells

Shoreline Management Plan (SMP) – a plan that takes into consideration the risks of coastal processes and attempts to identify sustainable coastal defence and management options

slab pull – when newly formed oceanic crust sinks into the mantle, pulling the rest of the plate further down with it

social clustering – a preference for people to live close to others they associate with, for example communities of similar ethnicity or religion

solar output – the amount of radiation that the sun emits (associated with sunspot activity), which can affect the Earth's temperature

solifluction – a form of *mass movement which is the downhill flow of saturated soil

Special Economic Zones (SEZ) – set up by national governments to offer financial or tax incentives to attract *Foreign Direct Investment, which differ from those incentives normally offered by a country

stadials – short-lived colder periods within a major glacial, associated with ice advance

stakeholders – people who have an interest in a scheme or an area, such as local residents or an environmental group

sub-aerial processes – the processes of weathering and *mass movement

subduction zone – the area in the mantle where a tectonic plate melts

sublimation – the change from the solid state to gas with no intermediate liquid stage

submergent coastline – when a rise in sea level floods a previously exposed coast

subsidies – grants given by governments to increase the profitability of key industries

supraglacial – on top of a glacier

swash-aligned – where sediment moves up and down the beach with little lateral transfer

swell waves – waves originating from the mid-ocean which appear as larger waves amongst smaller, locally-generated waves

T

tariffs – a tax that is paid on goods coming into or going out of a country

terminal groyne syndrome – when higher rates of erosion occur immediately after a set of coastal defences have finished

trade liberalisation – the removal of trade barriers such as *subsidies, *tariffs and *quotas

trade protectionism – the use of methods such as *tariffs and *quotas to attempt to boost a country's exports or reduce its imports

trading blocs – when countries have grouped together to promote free trade between them. The EU is an example of a trading bloc

transform fault – a fault created on a large scale when two plates slide past each other

tundra – *periglacial regions found in the barren plains of northern Canada, Alaska and Siberia where both temperature and rainfall are low

U

urbanisation – the increasing proportion of people living in towns and cities as opposed to the countryside

V

valley glaciers – large masses of ice moving from ice fields or *corries and following river courses

W

World Bank – a global organisation which uses bank deposits placed by the world's wealthiest countries to provide loans for development in other countries

world cities – cities with a major influence, based on: finance, law, political strength, innovation and ICT

World Trade Organisation (WTO) – a global organisation which looks at the rules for how countries trade with each other

Date: _____

	Revision Period 1	Revision Period 2	Revision Period 3	Revision Period 4	Revision Period 5
Monday					
Tuesday					
Wednesday					
Thursday					
Friday					
Saturday					
Sunday					

Date: _____

	Revision Period 1	Revision Period 2	Revision Period 3	Revision Period 4	Revision Period 5
Monday					
Tuesday					
Wednesday					
Thursday					
Friday					
Saturday					
Sunday					

Revision planner

Date: _____

	Revision Period 1	Revision Period 2	Revision Period 3	Revision Period 4	Revision Period 5
Monday					
Tuesday					
Wednesday					
Thursday					
Friday					
Saturday					
Sunday					